Iwígara

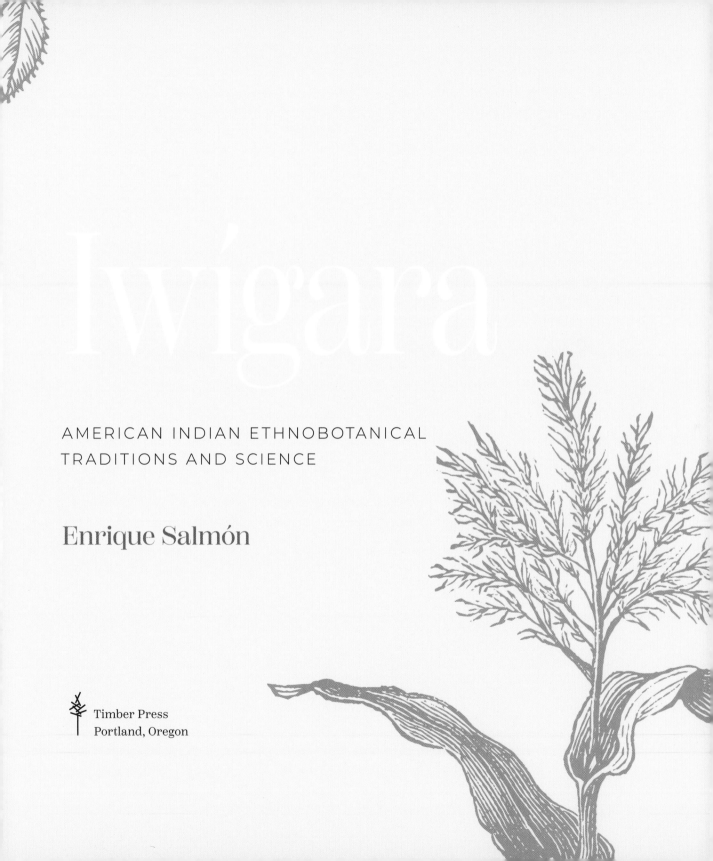

Iwígara

AMERICAN INDIAN ETHNOBOTANICAL TRADITIONS AND SCIENCE

Enrique Salmón

Timber Press
Portland, Oregon

Photography credits appear on page 231.

Cover illustration credits: (Front cover) Shutterstock: Andrenko Tatiana; AVA Bitter; Dn Br; Hein Nouwens; mamita; MicroOne; Morphart Creation; Nastasic; Olga Korneeva. Adobe Stock: Morphart Creation. (Back cover) Shutterstock: Morphart Creation.

Published in 2020 by Timber Press, Inc.

The Haseltine Building
133 S.W. Second Avenue, Suite 450
Portland, Oregon 97204-3527
timberpress.com

Printed in China on paper from responsible sources
Fourth printing 2022

Text design by Lauren Michelle Smith

Cover design by Faceout Studio

ISBN 978-1-60469-880-0

Catalog records for this book are available from the Library of Congress and the British Library.

To my grandmother Maria and my mother, Esperidiona.
You taught me so much about plants and about life.

Happily abundant passing showers I desire.
Happily an abundance of vegetation I desire.
Happily an abundance of pollen I desire.
Happily abundant dew I desire.
Happily may fair white corn, to the ends of the earth, come with you.
Happily may fair yellow corn, to the ends of the earth, come with you.
Happily may fair blue corn, to the ends of the earth, come with you.
Happily may fair corn of all kinds, to the ends of the earth, come with you.
Happily may fair plants of all kinds, to the ends of the earth, come with you.

NAVAJO (DINÉ) NIGHT CHANT

Contents

The Plants

Introduction

I was raised by a mother, a grandmother, and other extended family members who were living libraries of indigenous plant knowledge that has been collected, revised, and tested for millennia. This kind of knowledge is not housed in a building, stored on shelves according to some fixed system; it lives in the memories, oral literature, and daily rituals of its practitioners. As I grew up, I was exposed to this plant knowledge during regular plant-collecting excursions, and it was reinforced each time we cooked a traditional meal, every time my mother or grandmother treated me for an ailment, whenever I helped my grandfather in our cornfield. I took this early exposure to plant knowledge for granted, not knowing that it was becoming a rarity among most young people in our modern society.

Even as I was learning the practicalities of collecting, preparing, and administering plants, I was being exposed to a worldview that perceives plants as relatives. I was raised to respect all things, especially plants, as living beings. Now, as a native ethnobotanical scholar, I have struggled with how to translate this worldview into terms that can be understood by non-indigenous people. I am often asked, what is ethnobotany? My usual answer is to quote Richard I. Ford; he was one of my mentors, and he defined ethnobotany thus: "the study of direct inter-relationships between people and plants." Normally that response elicits blank looks. Perhaps it is better to tell how I followed my path into this field.

During my university years, whenever I was given the opportunity to choose a topic for a research paper, I invariably wrote about American Indians and plants. My master's degree thesis focused on medical plant knowledge and shamanism among three native cultures of the Southwest. It was at this point that my graduate advisor informed me that I was clearly on track to becoming an ethnobotanist.

Ever since that conversation I, like hundreds of ethnobotanists around the world, have devoted my life to studying how all people, not just indigenous peoples, interact with the plants in their local environments. It is painstaking and rewarding work that requires years of observing and participating with people in their local environments as they talk about plants, as they gather and use plants, as they make decisions about where to harvest plants, and as they sometimes battle with local authorities in order to protect plant-gathering areas and access to plants. I have participated in applied ethnobotanical work in Ethiopia, Australia, and North America, from Alaska to Mexico. I go, gladly, wherever native peoples are demanding their rights to sustainably manage their landscapes as their ancestors have done for thousands of years. In other words, ethnobotanists have become advocates for environmental justice, applied conservationism, and indigenous rights.

In 2000, I wrote an article for *Ecological Applications*, "Kincentric Ecology: Indigenous Perceptions of the Human-Nature Relationship." In it, I drew upon ideas and concepts from my own tribal worldview of our relationship with and responsibility to place, in an effort to explain how American Indians see themselves as part of an extended ecological family. I focused on the Rarámuri concept of *iwígara*, writing, "Iwígara channels the idea that all life, spiritual and physical, is interconnected in a continual cycle [and] expresses the belief that all life shares the same breath. We are all related to, and play a role in, the complexity of life."

In a worldview based on iwígara, humans are no more important to the natural world than any other form of life. This notion influences how I lead my own life and guides many of my decisions. Knowing that I am related to everything around me and share breath with all living things helps me to focus on my responsibility to honor all forms of life. I carefully consider all living and non-living things when making choices or weighing actions I might take. In short, I see myself as one of many stewards of the land and natural world. I share breath with it, so I endeavor to minister to it with appropriate ritual, thought, and ceremony.

Before writing this book, I conferred with native plant practitioners, my professional ethnobotanical network, and with close friends. I asked these knowledge holders and wisdom keepers to help me compile a list of plants that are the most culturally relevant to North American native peoples. Instead of trying to create a comprehensive list of every plant known to be used by one populace or another and all their potential uses, my goal was to focus on the plants that, overall, hold the most significance for indigenous peoples of the contiguous United States and Canada. In other words, what plants are the most important in the minds of native North American peoples? The answer is the following volume, consisting of 80 plant entries.

Ethnobotany is an area of study that interests both academics and laypeople. Unfortunately, most ethnobotanical texts are written by and for academics. The layperson, even one who is keenly interested in ethnobotanical knowledge, may not wish to decipher the esoteric jargon and not-so-engaging writing styles of professional ethnobotanists. Numerous books and other sources of information about American Indian plant knowledge focus on the same plants over and over, or read more like medicinal plant dictionaries. I tried to do something different here: to present a selection of American Indian ethnobotanical knowledge that is scientific in scope but written in an accessible style and—most important of all—to blend basic scientific and medicinal information with culturally specific knowledge and culturally relevant applications. I note, therefore, how some food plants, more than merely being sources of calories, play important cultural and ceremonial roles—for example, that wapato is not only an edible tuber but also a currency in gambling games, or that an elderberry's branches make good stems for smoking pipes. Although much of this knowledge is centuries old, I have, as much as possible, augmented it with updated contemporary information.

American Indian knowledge is most often transmitted through story. Story is the fabric of our humanity. Story structures our cultures. Story affirms and activates our identities. Each person wakes up to and lives their individual, cultural, and national story every day. Here, in these pages, I also include stories, myths, and narratives about plants that provide an avenue for the reader to learn more than their names, wildcrafting information, and uses. It is my hope that as the information related in this book is transmitted through story, the reader will be introduced to the spirit and character of these featured plants.

All Native Knowledge Is Local

The 80 plant entries in this book reflect knowledge from native tribes across North America. Throughout the book, I apply various modern identifiers to geographically and socially recognized regions of our continent; it should go without saying, however, that plants do not recognize abstract geopolitical boundaries. Many of the plants herein can be found outside of the areas specified in their entry. In addition, as a result of trade and periodic movements, native peoples were often familiar with plants that grew outside of their cultural boundaries. Sometimes knowledge of plants was transmitted in gender-specific channels, or a clan claimed "ownership" of certain plants. Some plants were reserved for use by the recognized healers or medicine people in a community.

   In my American Indian studies classes and whenever I am speaking to anyone who seems interested, I remind people that American Indian traditional knowledge is tied to the landscapes they called home. I often joke that it would be difficult to be a traditional Apache in Vermont: the landscape is too different. What's more, American Indians were never a homogenous culture. Over 500 languages were spoken in North America prior to European contact. Each one of those languages represented a distinct culture. Each of those cultures developed a lifestyle and library of ecological knowledge that fit the ecosystems and landscapes that they had sustainably managed and lived with for centuries. Although the ethnobotanical knowledge expressed in this book will sometimes overlap and be similar among indigenous cultures, in many instances the uses of plants will be representative of a specific culture's adaptation to an ecosystem and geographical region. Here I will give impressionistic sketches of these geographic and cultural regions, as they might be perceived by the people indigenous to them.

NORTHEAST

Native peoples from the Northeast tell this origin story. A long time ago, when the earth was only water, a certain young woman lived in the sky world, a sort of island of clouds floating high above the wet world below. She lived happily and comfortably with many other sky people, up there among the clouds. This particular woman discovered that she was pregnant with twins. For some reason, this news angered her husband, who flew into a rage. During his tantrum he approached a tree that grew tall in the sky world and tore it up by its roots. This created a hole in the ground of the sky world. The pregnant woman looked down into the hole and marveled at the water covering earth below. While she was peering through the hole, her husband pushed her. She plummeted down toward the waters.

It so happens that many water animals already existed on this wet earth. Luckily for the falling woman, two geese saw her tumbling from the sky. They flew toward the woman and managed to break her fall, capturing her on their soft wings. The geese gently deposited the woman onto the back of a big floating turtle, and from her perch there, the woman was introduced to other water animals, such as Beaver, Muskrat, Swan, and Osprey. The woman was grateful for her rescue but soon began to worry. How was she going to raise her twins on the back of a turtle?

The animals met and decided that, in order to create an island for the woman to live on, mud was going to have to be brought up from underneath the waters. Beaver tried to swim down to gather mud, but it was too deep and he soon ran out of air. Muskrat tried but ran into the same problem. It was just too deep. Other animals tried, without success. Finally little Toad dove into the water. He was gone for a long time, and the other animals became concerned. Then, all of a sudden, Toad floated to the surface. The animals thought he was dead, but when they looked more closely, they could see that not only was Toad still alive, but he had a mouth full of mud. The animals took the mud and spread it over the back of Turtle. The mud began to dry. As it dried, it spread—and continued to spread, until it formed what is now the ancestral lands of American Indians.

The woman moved about this Turtle Island, waiting for her twins to be born. While she waited she took dust from the new land and tossed it into the sky, creating the stars, the moon, and the sun. Eventually her twins were born. One she named Flint and the other Sapling. Flint, an evil twin of sorts, was responsible for bony fish, thorny bushes, and other bad things. Sapling was responsible for creating the rivers that teemed with fish, songbirds, edible and medicinal plants—all the things we love about our land.

The Northeast, from the Canadian Maritimes south to where the Delaware River spills into the Chesapeake Bay and east to the shores of Lake Erie, is a wonderfully lush region of mixed deciduous and conifer forests. The thick woods are interspersed with numerous creeks and rivers and open meadows that support a variety of useful grasses. Herbaceous plants occupy those places where meadows meet the shade of maples, birches, ash, red and white oak, white cedar, black walnut, and so many others. The forest understory includes myriad shrubs, more herbs, berries, fungi, and tubers.

For centuries native communities thrived in this landscape. They hunted the forests, fished the rivers, practiced small-scale agriculture, and gathered the bounty of their homelands. Many of the native communities of this region are Algonquian speakers; this family of languages is spoken by 29 tribal nations, including the Ojibwe, Abenaki, Potawatomi, Narragansett, Mohican, Cree, and Micmac. Other peoples of this region, such as the Seneca, Cayuga, Onondaga, Susquehannock, Mohawk, Oneida, and Huron, are Iroquoian speakers. All these peoples shared a version of the story just related regarding how the world began. There were other origin stories as well, but all tell how the animals emerged, and how the first people found their way to Turtle's back. No matter the version, these cultural histories are encoded with biocultural and ecological lessons that are woven into the daily fabric of indigenous activities. Included in many of the stories are the plants and trees of the Northeast landscape.

SOUTHEAST

Before they were forcibly removed from their homeland, the Cherokee were people of the Southeast, a diverse landscape shaped by hills, mountains, savannahs, marshes, and coastal scrublands. They shared this region with the Chickasaw, Creek, Seminole, Choctaw, Powhatan, and too many other tribes to name here. Like the Northeast, theirs was a land of abundance, a place where useful herbaceous plants kept company with pine, red maple, beech, ash, and walnut trees. The Cherokee knew that it was not always like this. In fact, they were latecomers to their lands. Their origin myth tells of how, in a distant past, the world was divided into three levels: the sky world, which is the upper level; the water world, which is the lower level; and the middle world, which is the earth. The animals occupied the sky world. They were curious about what lay below. So they sent little Water Beetle down to investigate. Water Beetle darted this way and that but could find nothing of significance. Finally, Water Beetle dove deep, until it reached the bottom of the water. It pulled some mud back up to the surface, where it began to dry and spread out, creating an island. So that it wouldn't float away, the island was attached to the upper level, where the animals dwelled, with four cords. The island remained wet and soft for a long time. Finally Buzzard flew down around the island to investigate. Buzzard flew low, and everywhere its wings touched the soft land, a valley or mountain was created, forming the land of the Southeast as we know it. The animals invited corn and other plants useful as food and/or medicine to occupy this new land. A human brother and sister followed the plants and began to multiply. Like so many native peoples' origin myths, agriculture plays a central role in this one. To native peoples, agricultural endeavors involve much more than knowing when to plow, how to irrigate, and at what depth to sow seed. The responsibility of growing food for one's community is connected to one's identity as a member of the community. This identity, this sense of being-ness, is tied to the history of the people on a landscape.

GREAT PLAINS

The Great Plains is buffalo country. Or, at least it used to be. Before the newly minted French and British colonists began to surge west, the Great Plains were shaped largely by the movement of sharp-hooved buffalo that tilled the soil and encouraged a distinct ecosystem, seed by seed on their shaggy coats and through their piles of dung. Over 50 different varieties of grasses once graced the plains. There were edible panic and tripsacum grasses, fescue, long-lived and iconic buffalo grasses, and the tall and flowing bluestem. The grasslands were broken up by swaths of willows, Osage orange, hackberry, and cottonwoods.

COMING TO TERMS

Each plant entry includes suggestions on how to process the parts of the plant for use as food and/or medicine. Below are the terms that are used most often throughout the book.

- Decoction: Plant material boiled down into a thick broth, syrup, or paste.

- Infusion: A solution made by soaking one or more parts of a plant in water over a period of time. The resulting liquid is then taken orally.

- Poultice: A thick paste made from one or more parts of a plant soaked in water. The resulting soft, damp mass of material is then applied externally in a thick layer to an injured part of the body, normally to treat soreness or inflammation.

- Wash: A single herb or blend of herbs infused or steeped in water. The resulting wash is then used externally on the affected body part, which is submersed or soaked directly in the wash solution.

To the north were the sacred Black Hills, covered with tall cedars and spruces that, in the minds of many native peoples, acted as protectors of their lands. According to the Lakota, the presence of these guardian conifers is not the only reason the Black Hills are sacred. It is there where "the people" emerged. A water monster had fought a battle with the people. The monster won and proceeded to flood the world. Everywhere there was water, except for a high hill where a lone girl survived. She was visited by a spotted bald eagle, whom she later married. The marriage produced two children, a boy and a girl. All this occurred in the Black Hills, and all Lakota people are descendants of these children. After the waters receded, what was left were the flattened plains and rolling hills of the Great Plains.

Many people of the plains—the Lakota, Cheyenne, Blackfoot, Kiowa, Comanche, Pawnee, and so many others—were nomadic. Others, such as the Omaha, Osage, Mandan, and Wichita, stayed in one place or were semi-nomadic; they traded their corn and other foods with those who chose to follow the herds. The nomadic peoples roamed large territories on horseback, becoming familiar not only with the movements and characteristics of the buffalo, antelope, elk, and other animals but also with the growth patterns of the plants that composed the luxurious grasslands. In the eyes of many early American explorers, the nomadic lifestyle of Plains Indian cultures epitomized individuality, freedom, courage, and nobility.

MOUNTAIN WEST

To the west of the Great Plains were the Rocky Mountains. The caretakers of the elevations and valleys of the Rockies and the Intermountain West were the Ute, Arapaho, Crow, Flathead, Shoshone, Jicarilla Apache, and Nez Perce. They indeed perceived themselves as caretakers. Their origin stories include morals that suggest they were chosen to occupy their mountainous environments in order to protect them. The southern Ute believe that curious Coyote was given a bag by Creator. Coyote was supposed to carry the bag—with its unknown contents—over into the mountains and leave it there. Coyote, ever sneaky and curious, traveled a distance and stopped to peek into the bag. As soon as he opened it, little creatures began to swarm out of the bag. They were people. Coyote managed to close the bag before all of the people could escape. The ones that got out of the bag ran away in all four directions. Coyote reached the mountains, where Creator told him to leave the bag. When he opened the bag, there were only a few people left inside. They were the people who became the dwellers and caretakers of the mountainous lands.

The people of the mountains were few in number but developed lifestyles that took advantage of what was offered by the seasons as well as by the different elevations. They knew how to use the different kinds of aspen, piñon, cedar, and dogwood for medicine, food, and for building shelter. They often stayed in the lower elevations in order to take advantage of mountain mahogany, chokecherry, currant, nahavita, and all the Rocky Mountain plants that have adapted to cold winters, short summers, and high elevations. They traveled east onto the plains in order to hunt buffalo and traded for foods with their Pueblo neighbors to the southwest.

COYOTE AND OTHER TRICKSTERS

In several plant entries in this book, Trickster figures play a role in how native peoples relate to plants. Trickster stories are most often equated with Coyote, but Coyote is only one of many such figures among indigenous peoples. The Trickster might be Raven or Rabbit. To my people, the Rarámuri, Skunk is often a Trickster. No matter the image, the concept of Trickster is reflective of an American Indian consciousness, or mindset, that embraces the gray and uncertain personality of the natural world. These parts of our universe are often not completely knowable and are, therefore, often considered sacred. Trickster occupies that gray, uncertain, sacred space. The Trickster brings to life the playful, disruptive, and creative side of human imagination. Re-creating is a process that native peoples look forward to over an annual cycle. Every ceremony, ritual, dance, healing, and initiation is an opportunity to participate in the ongoing creation of the universe and our relationship to it.

From disturbance emerges new birth and diversity. We should not fear that which is different or new. Through Trickster, we learn to embrace nonpolarity. Color blindness is assumed as well as every variation of gender. Therefore, Trickster expands the indigenous consciousness by freeing all constraints and creating an opening and threshold for flexibility and change. Through this kind of consciousness, culture and society are in a better frame for resilient thinking and adaptation. Blending the "old ways" with the new, then, is a virtue and a strength that is evident in American Indian cultures.

SOUTHWEST

The Hopi have occupied stone and adobe villages atop three arid mesas on the Colorado Plateau for more than a millennia. They believe Spider Woman led them up through a hole in the ground and into this world, the Fourth World, in order to act as caretakers of the earth. The previous three worlds were idyllic. The First World was destroyed by fire after the people became too corrupt and forgot to take care of the land. The Second World was destroyed by ice for the same reason. In the Third World, the people built great cities and developed weapons that could kill their enemies from great distances; but again, they did not care for the land, and their world was destroyed by floods. This was when Spider Woman led those who still remembered their caretaker duties into this, the Fourth World. The Navajo have a similar origin story but believe that we are residing in a Fifth World. The Tewa, who live along the Rio Grande River, north of Santa Fe, believe that they originated at the bottom of a shallow lake in southern Colorado; after their emergence, they migrated to where they now are.

No matter the story, native peoples of the Southwest (including the Southern Basin and Range) feel that they were led to their current homelands for a specific reason: to care for the land and to act as models of how to live properly with it. For thousands of years they have fulfilled this charge. Of all native peoples in the contiguous United States, the peoples of these arid regions have remained most admirably resilient, adhering to their lands, their languages, their spirituality, their food ways, and their plant knowledge.

The Southwest is a land of extreme contrast, host to an incredible array of biocultural diversity. These words may sound like a clichéd travel brochure, but they express an absolute truth. Heading south, it is possible to drive, in one day, from the pine forests and upper sagebrush landscape near Durango, Colorado, through the dry, red Colorado Plateau chaparral, through some more pine forest near Chama, New Mexico, to the cacti and yucca of the Chihuahuan Desert near the Mexico–U.S. border. Alternatively, beginning again near Durango but heading in a more southwesterly direction, it's possible to pass through the arid plateau just south of Monument Valley and make a left near the pine-forested region around Flagstaff, Arizona, to wind up in one of the most biologically diverse regions in the world: the Sonoran Desert. In any direction, you would drive over the ancestral and current homelands of indigenous communities representing five to nine separate linguistic and cultural groups.

Up on the Colorado Plateau the Hopi continue to practice the Hopi Way, a spiritual lifestyle that does not strive for a specific outcome or product but rather is a journey, focused on what is learned along the way about their relationship to place and community. It is a way of resilient persistence; the Hopi have been able to successfully endure shocks to their social and environmental systems. The same can be said of most of the Pueblo peoples; not only the Tewa but also the Acoma, Zia, and Taos originated in other places but, in response to environmental changes, migrated and adapted to a new situation. The Apache came from parts far north but, after continued and sometimes violent movement, ended up in the Southwest and adopted a nomadic lifestyle. Even Hispano peoples who eventually *became* indigenous to the Southwest mixed traditions that they brought with them from Europe with those of their indigenous neighbors, creating the acequia system of sustainable agricultural practice, for example, which has created new highways of ethnobotanical diversity and a source of useful plants for the people. All these examples are located in specific landscapes across the Southwest. Each location is an example of cultural and ethnobotanical refugia and resilient culture and practice; each multigenerational community maintains a deep sense of connection to the arid landscapes, which reaffirms long-term memory loops concerning traditional land management and identity.

PACIFIC NORTHWEST

Lewis and Clark's Corps of Discovery stumbled into the Pacific Northwest in 1805; they had already survived a harsh winter, a difficult crossing of the Continental Divide, dysentery, and a descent of the Columbia River. When they finally arrived at the coast, they were faced with unending rain, seemingly impenetrable forests with trees taller than any they had ever seen, and native peoples who were extremely comfortable in this landscape and who were indifferent to the newcomers.

The Pacific Northwest is a lush landscape populated by a similar abundance of peoples. There are the Tlingit, Haida, Tsimshian, Kwakiutl, and Bella Coola, the Nuu-chah-nulth, Quileute, Coast Salish, and Chinook. A little south along the coast of what is now Oregon lived the Tillamook, Siuslaw, Umpqua, Yurok, and Hupa. These people spoke diverse languages from the Salishan, Wakashan, Chimakuan, Penutian, and Athabaskan linguistic families, but they lived similar lifestyles. They took full advantage of a temperate climate; the Pacific Ocean current, although it did carry lots of moisture, also encouraged seven kinds of salmon that spawned in the many rivers. Dense coniferous forests supported large mammals and a variety of useful food and medicinal plants.

The current landscape and forests are not that different from what they were in 1805. Unfortunately, overfishing and dams have reduced wild salmon populations, and clear cutting and mismanagement have negatively affected plant and animal life. However, Pacific Northwest peoples persist: a number of years ago, while traveling along the shores of Puget Sound, I stopped at a Skokomish community for a rest. I struck up a conversation with a young man—one of a group of several young Skokomish who had had to convince their elders to teach them how to build large ocean-going canoes, so that they could participate in the traditional canoe races held by other native communities. These canoes were central to the traditional livelihoods and cultures of the Skokomish and other coastal natives. What the young people had not counted on was that the process of building a canoe would lead to a resurgence of their language and relationship to home landscape.

Canoe building is more than gathering materials and putting them together properly so that it all results in a vessel that floats. The young Skokomish were first taught the proper cedar trees to harvest, how to harvest them with respect and ritual, when to harvest, and which songs to sing in the process. They had to perform these rituals in their native language. They were told stories that related to canoes and cedar trees. They learned about the ecosystems that the canoes were harvested from and how they, the young canoe builders, were directly related to it. The young man told me that when they began the process of learning how to build a canoe, their language was almost gone and the people no longer felt a strong indigenous identity. By the time I met him, their language was resurgent, rituals and ceremony were again being performed, and they were involved in sustainably managing their local ecosystem.

WEST COAST

Before the arrival of Europeans, the West Coast from British Columbia south into Baja California was the most diverse cultural area in North America. It is estimated that more than a third of all native peoples of North America lived here, in yet another region of abundance. Visitors still flock to the coast and sierras to marvel at the giant sequoias and redwoods; the arid regions and deserts that lie in the shadows of the Sierra Nevada are home to unusual (and frequently endemic) plant life. The Central Valley of California is often referred to as America's breadbasket, but before the arrival of European agriculture, native peoples of the West Coast were unfamiliar with this concept: they had not had to assume a sedentary agricultural lifestyle in order to secure their sources of plant foods. The numerous species of oaks provided a seasonal crop of acorns, which was supplemented throughout the year by other nut and fruit trees, tubers, chenopods, berries, and herbaceous plants. Rivers and streams teemed with fish, and there was plenty of game for the people to hunt.

A commonality of this region is that the lands and waters were created by hero-beings that battled monsters and other creatures on behalf of the people. Often, it is Coyote who either intentionally or clumsily played a role in forming

RESPECTFUL HARVESTING

A worldview based on iwígara compels me to act as a responsible steward of my plant relatives. It reminds me also to always give thanks to the plants when I am gathering and collecting. I will make an offering and verbally offer my gratitude to the plants being collected as well as to the land. I am also careful to never overharvest and to spread my harvesting among several plants and locations. I am not suggesting a specific method for your own collecting; I am merely suggesting that you keep these thoughts in mind if you choose to wildcraft plants that are highlighted in this book.

the lands of the West Coast. And not only that: the Miwok believe that Coyote played a direct role in bringing people into existence and forming how human beings look today. The Yokuts believe that Coyote was helped by Eagle when the first humans were formed from clay. The Salinan think that Eagle made First Man from clay and then created First Woman from one of its feathers.

Over 100 distinct languages, separated into more than 300 dialects, were spoken across this landscape. The languages were derived from several linguistic families, including the Yukian, Maiduan, and Uto-Aztecan groups; Esselen and Karuk are linguistic isolates, with no connection to any other language group. Each language of this region represents a culture with its own worldview and library of ecological knowledge. But all native peoples of the West Coast engaged in some form of complex and sophisticated "gardening" of their homelands, whether grassland, mixed woodland, wetland, chaparral, or conifer forest. Their systems of land management—seasonal pruning, coppicing, and low-intensity fire regimes—not only ensured they had what they needed in terms of food, medicinal, and utilitarian plant materials but also encouraged plant and animal diversity. These repeated cycles of clearing, fire, and careful use, on a scale unimaginable today, may be considered a form of advanced permaculture.

Much of the pre-Columbian landscape of the West Coast has been transformed over the last 500 years, but if one knows where to search and what to look for, most of the native plants can still be found. In addition, despite the near genocide of the native communities of this region, many indigenous peoples have persevered and remain resilient on their ancestral landscapes.

The
Plants

ASH

FRAXINUS SPP.

Family: Oleaceae
Parts Used: twigs, bark, wood
Season: year-round
Region: Northeast

"Choose the right tool for the job," the old maxim has it. Similar thinking can be applied to selecting woods for harvesting and carving into specific objects. Every type of wood has its own characteristics, advantages and disadvantages. Sometimes a wood offers a seemingly unlimited variety of uses; ash is such a one. Ash trees offer a versatility of purposes to American Indians not found in many other plants.

USES

I have been told several times that tools and other implements constructed of ash seem to have a spirit, or to be alive. I cannot fully

A mature *Fraxinus americana* on the University of Connecticut campus.

explain why others would sense this about ash, but I can speak from my own experience of it as a woodworker. When ash wood has been properly seasoned and milled, its grains seem to let the woodworker know which way and in what manner to carve, cut, or plane it. When completed, tools made of ash are lightweight yet durable, able to withstand abrasions and knocks with little wear and tear; and they are flexible when they need to be. The wood is also strikingly colored, a shade of yellow infused with turmeric and saffron, and the exposed grains and knots of the wood add visual interest.

For millennia, ash has been made into canoe paddles; they remain tough even when water-soaked and can take being knocked around by rocky river bottoms and shorelines. Ash can also be steamed and bent into snowshoes and frames for cradleboards, or heated and split into small thin splints for making baskets of varying sizes. Ash has been made into durable lances as well as the straightest arrows and solid bows. Birch bark canoes are framed by split and bent ash. As these and many more examples suggest, ash has been a preferred wood for native peoples; however, other parts of the tree are also put to good use.

The inner bark of ash trees has many medicinal uses; that of green ash (*Fraxinus pennsylvanica*) can be scraped and made into an infusion for fatigue and depression, or be used as a diuretic and a laxative against worms. It can also be chewed and applied to skin sores. The inner bark of white ash (*F. americana*) has been used to help stimulate menses and as a liver and gastrointestinal tonic. The twigs and inner bark of Oregon ash

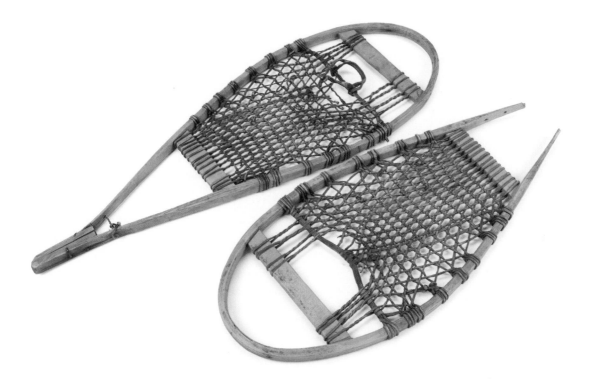

Nipmuc snowshoes made of ash, hide, and nails, circa 1790.

(*F. latifolia*) are used to reduce fevers and as a snake repellent. Other species of ash offer similar medicinal uses.

IDENTIFICATION AND HARVEST

If you were to ask the average person to name an iconic tree you would receive all sorts of answers, depending on where the person lives or grew up: a Navajo in northern Arizona will respond differently from a Nipmuc in Connecticut. Most of the time, my students respond with pine, maple, or oak. Interestingly, I have never had a single person name the ash. Yet, ash trees are ubiquitous throughout North America. About 20 species and subspecies of ash have been identified throughout the continent; this number does not include those introduced from Europe, Africa, and Asia.

Ash trees grow 30 to 60 feet tall. They are deciduous, except for a couple of species that grow in more tropical regions. The leaves are elongated-oval, with fine teeth at the edges. They are opposite and pinnately compound,

A 19th-century botanical illustration of *Fraxinus americana*.

on female trees) are elongated, greenish, and winged, about 2 inches long; they begin to drop from the trees in late summer, slowly spinning their way to the ground like little helicopters. In the fall the leaves turn a rusty yellowish orange. Bark is thick and grayish; during a tree's early years it is relatively smooth but begins to develop fissures throughout the trunk that deepen with age.

Most native peoples recognized that useful woods of deciduous trees are best collected from late winter to early spring, just as the earth is beginning to push its life-sustaining saps back up through the roots and into the trunk and branches. If the wood is collected in the dead of winter, it will not be very useful because some of the moisture in the wood is still present and frozen. As the wood thaws, the moisture exits the wood, leading to brittleness. If it is collected later in spring through fall, the wood still holds too much sap and moisture; wood harvested at this time shrinks as it dries and tends to crack and splinter easily. Collect wood in late winter or early spring, when the sap/moisture to wood ratio is just right: the wood will be strong, resilient, flexible—essential characteristics given ash's various traditional uses.

HEALTH BENEFITS

Fraxinus species contain secoiridoids, phenylethanoids, flavonoids, coumarin, and lignans. These beneficial phytochemicals are anticancer, anti-inflammatory, antioxidant, antimicrobial, and neuroprotective, among other pharmacotherapeutic effects.

with leaflets in as many as nine pairs; the lower sides of the leaves tend to be lighter than the upper sides. In early spring, a cluster of very small purplish (sometimes greenish white) flowers sprout at the tips of the leaf areas, near the ends of the branches. Fruits (only

BEANS
PHASEOLUS SPP.

Family: Fabaceae
Parts Used: fruit
Season: summer, fall
Region: North America

Nearly every indigenous language in the western hemisphere has a word for beans, which is a reflection of the ubiquity of *Phaseolus* species in the Americas. Beans are second only to corn as the most consumed traditional food among American Indians.

February gets cold on the high Colorado Plateau of northern Arizona. Hopi communities are situated here, across three mesas. For the Hopi, February is the time of Powamuya, when the annual Powamu, or bean dances, are celebrated. It is also the time on the Hopi ceremonial calendar when the *katsinam*, or rain and earth spirits, return to the Hopi mesas in order to help the people prepare for the coming growing season and to oversee the initiation

A kiva at Mesa Verde National Park in Colorado.

Sorting a variety of dried beans.

of young Hopi into a Kiva Society. Kivas are subterranean structures where the different spiritual societies hold meetings and perform ceremonies. The Powamu ceremonies are actually several rituals spread over 16 days. The rituals honor the actions of fertility as well as germination.

During Powamu, beans are sprouted inside different kivas. The sprouting beans honor the coming growing season and are seen as a good omen for the success of the harvest. On the final day of the ceremonies, many katsinam gather and give away gifts to those in attendance. The gifts include lightning sticks, dancing sticks, arrows, rattles, moccasins, katsina dolls, and sprouted beans.

My people, the Rarámuri, call beans *muni-ki*. Among the Yaqui the word is *muni*, for the Hopi, *mun*. The similarity in nomenclature is because the three languages represent branches of the Uto-Aztecan linguistic family tree. The Lakota refer to beans as *oh-min'-nee-chah*; the Cherokee, *thuya*.

USES

I cook up a pot of beans nearly every week. Normally I cook pinto beans, but sometimes it's big, beefy, sweet runner beans or small earthy-tasting tepary beans. Or perhaps some yellow beans. Maybe I will do the pinto-like beans, white with splotches of black and brown, that resemble little black and white goats. The choices are many, and they all taste good. Across native North America, dried beans are cooked and eaten plain or with different spices and additional ingredients. They are main characters in soups and stews. They are cooked and mashed to make a soup base. The Cherokee will mash cooked beans and make them into bread or just add them to hickory-nut soup. Many tribes enjoy cooking beans with fresh and dried corn. It is likely that the original Thanksgiving feast included deer, water fowl, turkeys, shellfish, eels, squash, corn, and beans. Many American dishes that are cooked and eaten today have their origins in American Indian cuisine, including succotash, which combines beans with cracked corn. Many people still enjoy sweetened baked beans; and Tex-Mex, like so many meals in the Southwest, would feel empty without the ever-present side of frijoles.

IDENTIFICATION AND HARVEST

Today when we talk about beans, we are referring to one of two species. The first is *Phaseolus vulgaris*, an erect herbaceous bush, 1.5 to 5 feet tall, with many pubescent, willowy branches. The compound leaves are trifoliate with ovate leaflets, and the classic pea-shaped flowers, often white to pink, grow in clusters between the leaf axils. The fruit pods are elongate containing from three to ten beans. The pods are usually green but come in several colors; the beans also come in different shapes and colors. The second is *P. coccineus* (runner bean, pole bean), which is a perennial climbing vine growing up to 15 feet long. Unlike *P. vulgaris*, it has a tuberous taproot. The thicker stems tend to be prostrate for most of their length and rise toward the terminus. The trifoliate leaves emerge at the end of long petioles; the leaflets are 2.5 to 7 inches long and 1.5 to 4 inches across. The terminal flowers appear in pairs and come in white, purple, and red. Each flower produces a single pod, which is green, yellow, black, or purple; some are even striped. The pods can be somewhat knife-shaped, cylindrical, flat, or straight, 0.5 to 1 inch wide and up to 8 inches in length; they contain four to 12 large kidney-shaped seeds, in a wide range of colors: white, red, green, tan, purple, gray, or black.

The young immature pods of both species can be harvested and eaten while the beans inside are still soft. Most American Indians wait until the pods have dried on the bush or vine. The entire plant is collected and piled on top of blankets or tarps among the other harvested bean plants. The harvested plants are allowed to dry a little longer before they are shaken or beaten with brooms or sticks, which causes the dried beans to separate from the pods onto the tarps. The beans are collected and stored for later use.

Native peoples did not look to beans for medicine; however, recent studies confirm that the beans included in the ancestral diets of many American Indians played an important role in preventing nutrition-related disease, especially type 2 diabetes. Beans are high in fiber (soluble and insoluble), complex carbohydrates, and mucilage. Together these compounds help to control glucose and insulin levels and help any regular eater of beans to avoid obesity and high cholesterol. Leguminous fibers found in beans seem to control postprandial glucose levels, the sudden rise of which affects diabetics. An additional benefit of eating beans is the trace of chromium found in their outer coats; chromium deficiency has been found to cause insulin resistance. Beans are also high in protein, and their mucilaginous outer coats slow down the digestive process.

BEARBERRY

ARCTOSTAPHYLOS UVA-URSI

———

KINNIKINNICK

Family: Ericaceae
Parts Used: whole plant
Season: spring, summer, fall
Region: North America

Like many people, I am generally reluctant to say the scientific names of plants aloud. I have worked so long as an ethnobotanist that I recognize most of them, but my tongue tends

The bright red fruit of bearberry.

to trip over many of the unfamiliar (and often Greek) syllabic combinations. However, I have always enjoyed voicing bearberry's binomial, especially its specific epithet, *uva-ursi*. It just rolls off my Spanish-speaking tongue: its literal Spanish translation is "grape-bear," or, roughly, "grape of the bear." In my mind, the name conjures up happy-go-lucky bears rolling around in the sun in open pine forests, feasting on the little red berries of this plant. But there's more to the origin of the name. According to an Anishinaabe legend, after Bear and Otter pushed the first tree of life from the underworld up into the surface world, Bear was curious about the newly unfurled animals living on the surface. When he approached a group of them, they all scattered, leaving only an infant behind. Bear studied the infant, noticing that

Bearberry's delicate pink flowers in late spring.

it did not have a berry hanging down in the back of its throat. Bear knew that this part of his anatomy helped to keep all the berries that he consumed from coming back up. So Bear looked around and found a short mat of vine-like plants with shiny green leaves and red berries. He plucked one of the berries and put it into the baby's mouth. In so doing, Bear ensured that the people would be able to keep their food down and not starve. And that is only one of many uses for bearberry.

USES

Kinnikinnick (from the Algonquin for "smoking mix"), an alternative common name for bearberry, has been universally applied to all American Indian smoking mixtures since the Colonial era. An important trade item, the plant was known to all by this Algonquin name. Bearberry has long been used by American Indians across the continent as part of a smoking mixture that includes sumac, deer tongue, dogwood, and tobacco. My tribe, the Rarámuri, use an infusion of bearberry leaves most frequently for urinary tract infections. The same infusion can be applied to the skin to relieve itching, sores, cuts, and rashes. An infusion is used as a mouthwash for sore gums and cankers. The Cheyenne make an infusion of the entire plant and take it for back pain; they also make a poultice of the leaves and apply it externally for pain. The Cree mix the stems of bearberry and blueberry into a decoction to prevent miscarriage and to speed postpartum recovery. The roots are used to slow down menstrual bleeding. The Navajo drink a tea of the leaves as a ceremonial emetic. The Ojibwe take an infusion of the leaves to treat rheumatism; the Okanagan use a decoction of the leaves to treat sore eyes.

Many tribes across North America have various ways of eating the little mealy berries. They are added to soups and stews, pounded and mixed with dried meats, and are eaten raw or cooked. There is not much flavor to the berries when eaten raw, but they sweeten when cooked. When used as part of a ceremonial smoking mixture, bearberry has a mild narcotic effect.

IDENTIFICATION AND HARVEST

Bearberry is a short to prostrate woody shrub that often creates dense stands no taller than 6 inches but can sometimes cover over 100 square feet of ground. The evergreen leaves are simple, arranged alternately on short stems; they turn red in winter. Clusters of small pinkish to white urn-shaped flowers appear in May and June. The glossy red fruits that follow are about a quarter-inch in diameter and persist on the plant into winter. Bearberry is a hardy plant, tolerant of both cold and drought. Look for it in open, sunlit and shaded forested areas. It does well in disturbed habitats and grasslands, and on mountain slopes, plateaus, and sandy plains. In the Mountain West, I have often come across dense stands of bearberry at the edges of forests in sunny areas. The green leaves can be collected any time during the plant's growth period, from spring to fall. Look for the berries beginning in late summer into the fall.

Bearberry leaves contain hydroquinone, arbutin, tannins, quercetin, and myricetin. Hydroquinone and arbutin are antiseptic and antibacterial; they are especially effective against the bacteria associated with urinary tract infections. The tannins and quercetin are astringent and diuretic; myricetin is an antioxidant.

BIRCH
BETULA SPP.

Family: Betulaceae
Parts Used: bark, wood, gum, sap
Season: spring, summer
Region: North America

There was very little sound as we glided the birch bark canoe along the water. We could listen to the breezes and the bird song in the trees—undisturbed by our paddling. It was

A traditionally constructed birch bark canoe.

my first time in this type of vessel. The canoe belonged to my friend Bob Erb; it was hand-made for him by a master canoe builder who used the same materials as pre-Columbian Indians would have used. It was one of those rare utilitarian works of art.

The North American birch tree is more than a source for canoe-building material; it also provides materials for basket making and other crafts. Parts of the tree can be eaten. Other parts are medicinal.

USES

According to Abenaki herbalist Judy Dow, white birch bark is valued for its antifungal and antibacterial properties, which is why it was (and still is) used for food caches and baskets. It's also waterproof, hence its use for covering lodges and canoes. Judy also told me about chaga (*Inonotus obliquus*), a mushroom that is parasitic on birch and other trees. It appears as a burnt, cracked black growth that seems to sprout from between the seams of the birch bark; the inside of the fungus is brown-orange. The chaga mushroom is col-lected, dried, and ground into a powder, which can then be made into an infusion for gastritis, ulcers, to boost the immune system, and for inflammation.

A little further south, the Delaware and other peoples indigenous to Maryland use the bark of gray birch (*Betula populifolia*) to treat infected and swollen cuts. A decoction of the bark of paper birch (*B. papyrifera*) is used to treat dysentery, blood diseases, to induce sweating, and to ensure an adequate supply of

The papery white bark of a birch.

milk in nursing mothers. Modern manufactures derive xylitol, a sugar substitute, from the chewable gum of some birch trees. The gum is also a disinfectant and is said to afford a mild "buzz."

Some birch trees (e.g., sweet birch, *Betula lenta*) are tapped for their sweet sap. The boiled-down sap is thicker than maple syrup and tastes a bit more like molasses. River birch (*B. nigra*) can also be tapped for its sap; it's not as sweet as the sap of sweet birch, but it's often used for making birch beer. The wood of all birch trees, soft but finely close-grained and straight, lends itself well to being carved into bowls and other crafts. The peeled bark has long been used for constructing baskets, buckets, and storage bins, and, of course, canoes.

The antifungal properties of birch bark help to prevent anything that is stored in bins made of birch bark from being infested by pests. In regions where birch trees are abundant, the bark has also been used for the outer covering of sweat lodges.

IDENTIFICATION AND HARVEST

About 18 species of birch are native to North America. These deciduous trees are a bright note in the hillside mosaics of autumn color in the Northeast and other regions. A key identifier of the genus is the paper-like bark, which ranges from white (e.g., *Betula papyrifera*) to gray (*B. lenta*) to light brown to dark gray (*B. nigra*). Most birches have trunks that

An Athabascan birch bark basket, circa 1910.

Birches are surprisingly rich in phytochemicals. The gum, such as that from sweet birch, is a disinfectant; it contains xylitol and other medicinally active terpenes. The inner bark contains ten phenolic compounds, which together are immunomodulatory, anti-inflammatory, antimicrobial, antiviral, antioxidant, antidiabetic, antiarthritic, and anticancer, as well as being gastroprotective, hepatoprotective, and dermatoprotective.

are marked with irregular scales, plates, and fissures. The leaves are alternate, ovate to wedge-shaped, ending in sharp points; they are 2 to 3 inches long and usually serrate. The yellowish brown flowers are unisexual, growing as separate male and female catkins on the same tree. Some species of birch live only 20 years and reach a height of 30 feet; sweet birch and others can live over 300 years and get as tall as 115 feet.

Birches grow best in colder habitats among other deciduous trees, such as maple and ash. Collect birch bark in the spring when the sap is still running. If the bark is collected any later, the inner cambium of the wood will come off with the bark, which harms the tree. Using a very sharp knife, make a vertical cut, but only as deep as the cambium. Do not cut into the cambium. Make similar cuts around the tree at the top and bottom ends of the vertical cut. The bark will easily peel away from the trunk. Tap the sap of sweet birch about a month after maple trees are tapped; the sap will run faster than maple sap.

A 19th-century botanical illustration of *Betula lenta*.

BLACKBERRY

RUBUS SPP.

Family: Rosaceae
Parts Used: leaves, stems, fruit, roots
Season: summer, fall
Region: North America

At least once a year, out in my garden, I will be forced to dig up an errant blackberry sprouting underneath the California laurel or weaseling its way through large squash leaves. I am not sure how the blackberry plants get to my garden. The most likely culprits are the many birds that visit the garden—I have even spied blue jays burying peanuts in the soft soil of my planting beds, so nothing surprises me now. However their seeds arrived, if I didn't make the effort to uproot the invading blackberries, it would not be long before a section of my garden would be transformed into a nearly impenetrable bramble of dark green leaves and sweet berries protected by bloodletting thorns. The birds would be happy. I could leave one of the plants, with the best intention of maintaining and containing its size with constant pruning, but the blackberries that take over my garden are the nonnative Himalayan blackberry (*Rubus armeniacus*). Either way, it is an invasive blackberry. If one of the species of blackberry native to North America was growing in my yard, I would gladly welcome it; these have long been a favorite forage food for American Indians. And, in addition, wild native species offer more anthocyanins and other beneficial polyphenols; this is because domesticated varieties are bred to contain more water, which

Blackberries in various stages of ripeness.

leaches away the healthful compounds. Also, wild berries produce more phytochemicals as they work to fight off natural diseases, insects, and competition from other plants.

USES

The sweet fruit that we commonly think of as the "berry" of blackberries is actually a drupelet, an aggregate of small fruits. Native peoples across North America forage and eat them fresh, ground with wild game meat to make pemmican cakes, or dried and stored for later consumption. But blackberries are more than a sweet treat. They are useful in the treatment of cancer, dysentery, diarrhea, whooping cough, colitis, toothache, anemia, psoriasis, sore throat, mouth ulcer, hemorrhoids, and minor bleeding. Among many indigenous nations in eastern North America, an infusion of the roots, sometimes mixed with blackberry leaves, is taken to treat diarrhea and rheumatism. The Iroquois use the same infusion for colds and cough. The Menominee use an infusion of the roots as a wash for sore eyes. The Cherokee drink a decoction of the bark of blackberry stems for urinary tract infections. Abenaki herbalist Judy Dow weaves baskets with the stems after she removes their thorns.

IDENTIFICATION AND HARVEST

Blackberries are often confused with raspberries, a rose family relative. The difference is that when a raspberry fruit is picked, the torus (the receptacle at the end of the short stem on which the fruit rest) stays on the plant, leaving the characteristic hollow inside raspberry fruits. When a blackberry is picked, the torus stays inside the fruit. Blackberries are a perennial plant that produces several stems, or canes, from its root system. A stem can reach over 20 feet long and is often entangled with several other stems, creating thickets or brambles. The stems bear large, palmately compound leaves of three to five leaflets—and numerous short-curved, sharp prickles or spines. They are sharp and strong enough to rip through denim and even into leather gloves. White to pink flowers emerge in late spring and early summer on short racemes. Each flower is 1 to 1.5 inches across, with five petals.

Blackberries are one of the most widely distributed plants worldwide, with approximately 375 species, subspecies, and hybrids. A large

The flower of *Rubus ursinus*, a western North American native blackberry.

percentage are native to North America, occupying a diversity of habitats, from arid lands in the Southwest to the damp, windy coastal bluffs of California and frigid forests of Alaska. Plants are extremely successful at adapting to local habitats. Their seeds and root systems can lay dormant for several seasons before the conditions are right for them to sprout. They easily adapt to low light or direct light, changing soils, and availability of moisture. The seed coats are durable, which has encouraged their dispersal by birds throughout North America.

As with so many plants, the common names of blackberries often reflect the habitats and regions to which they are endemic or honor the western botanist who first identified them—for example, Andrews' blackberry (*Rubus andrewsianus*) and the rare Santa Fe raspberry (*R. aliceae*), the latter endemic to a single small county in north-central New Mexico; Lookout Mountain blackberry (*R. trux*), found in West Virginia, Georgia, Tennessee, and Kentucky; and Adirondack blackberry (*R. lawrencei*) of the northern Adirondacks.

HEALTH BENEFITS

Blackberry fruit is a great source of vitamin C and other important nutrients, and the berries contain numerous phytochemicals, including ellagic acid (an antioxidant), anthocyanins, and other polyphenols. The roots and leaves of *Rubus* species contain antibacterial flavonoids, salicylic acid, and myricetin.

BLACK COHOSH
ACTAEA RACEMOSA

BLACK BANEBERRY, BLACK SNAKEROOT, RATTLEROOT, BLACK BUGBANE

Family: Ranunculaceae
Parts Used: roots
Season: fall
Region: eastern North America

American Indian plant knowledge is sophisticated and complex. I always make this point in my classroom and written lectures. I note how Europeans and their descendants reached the shores of North America only about 400 years ago, and European-trained botanists really did not seriously begin to study the plants of this continent until about 200 years after that.

In contrast, if we assume the more conservative estimates for American Indian occupation of this continent, native peoples have been studying, observing, and using the plants of North America for at least 20,000 years—and I believe that number could actually be doubled. Whatever the number, that's a lot of time, over the course of which native peoples developed quite a sophisticated plant library and pharmacopeia. That library includes many uses for this versatile North American native plant.

USES

It should first be noted that black cohosh is toxic in large doses. It can irritate nerve centers and can cause abortions. Black cohosh can also cause gastrointestinal problems, nausea, and headaches if used improperly.

The intensely colored leaves and stems of black cohosh.

Black cohosh is valued by several eastern tribes, including the Cherokee, Delaware, Iroquois, Micmac, and Penobscot, as a general tonic, cold remedy, diuretic, sedative, dermatological and gynecological aid, abortifacient, antirheumatic, and analgesic. The Cherokee use an infusion of the root for hives; as an analgesic, antirheumatic, laxative, and sedative; and to treat colds and coughs, stimulate menstruation, and relieve cramps during childbirth. The Delaware include the root in a general tonic. The Iroquois use a decoction of the root as a soak to treat rheumatism. The root is also used as a blood purifier and to promote the flow of breast milk. The Micmac and Penobscot use the root as a kidney aid.

IDENTIFICATION AND HARVEST

Black cohosh is an herbaceous perennial that can grow up to 5 feet tall. It is characterized by a central erect stem that produces flowers from July to October. The flowers have no petals or sepals but occur as many tight long stamens surrounding a white stigma. Basal leaves are compound, with three leaflets. The root system is rhizomatous and not very deep into the soil. Black cohosh roots are typically collected in fall; this allows the plant to first reach maturity and drop its seeds. Also, during the fall, the roots contain less moisture, which leads to easier drying and storage. Look for black cohosh along the edges of moist mixed deciduous forests, on slopes, along creeks, and in moist meadows. It will grow at lower elevations in the Northeast and south into Georgia.

HEALTH BENEFITS

Black cohosh has three main active constituents: triterpene glycosides (including deoxyactein), flavonoids, and aromatic acids. Triterpene glycosides may connect with estrogen and serotonin receptors, which would explain the use of this plant for gynecological problems. All three flavonoids present in black cohosh root (biochanin A, formononetin, and kaempferol) can affect phytoestrogen activity and are antifungal, analgesic, and antibacterial as well. Black cohosh root's aromatic acids (including ferulic acid) are antioxidant and anti-inflammatory; ferulic acid is also an antispasmodic and sedative.

BLACKROOT
VERONICASTRUM VIRGINICUM

CULVER'S ROOT, BEAUMONT'S ROOT

Family: Plantaginaceae
Parts Used: whole plant
Season: spring, summer, fall
Region: eastern North America

Purification and cleansing, both physical and spiritual, are central to many American Indian rituals and lifestyles. This is why sagebrush, sweetgrass, and certain other plant-based smudges are burned in ritualized areas during ceremonies. However, the body must also be periodically purified and cleansed of unwanted elements. The Seneca, Delaware,

Blackroot in flower.

Choctaw, Cherokee, and other eastern tribal peoples would drink an infusion of blackroot in order to induce cleansing vomiting and purging prior to and/or during specific ceremonies. The beautifully erect and luxuriant blackroot plant is also a powerful medicinal for several ailments.

USES

Blackroot is used medicinally by several American Indian tribes to treat a variety of ailments. An infusion of the root of blackroot is taken as an analgesic to relieve backaches and rheumatism. A stronger infusion is used as a cathartic and emetic. It is also used as a diaphoretic, a disinfectant, and to relieve fevers. The Chippewa take blackroot to cleanse the blood, while the Iroquois take the root to relieve diarrhea. Blackroot is also taken as a laxative and to treat coughs and chills.

IDENTIFICATION AND HARVEST

Blackroot is an erect perennial that quickly grows up to 4 feet tall. Blackroot normally grows as bunches of individual plants. The round central stems are smooth, supporting whorls of three to seven serrated leaves that grow to about 6 inches. From July to September, candelabra-like inflorescences sprout from the central stem, bearing blue to pink

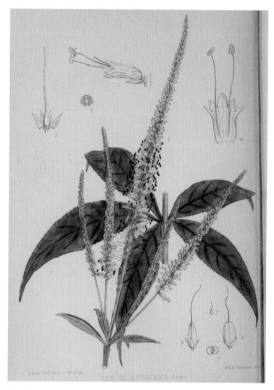

Blackroot resembles veronica but can be separated by its leaves, which grow in whorls, as seen in this 19th-century botanical illustration.

and whitish flowers; the inflorescence can grow to be 9 inches in length. The root system is characterized by a central taproot. Blackroot is found from the Canadian Maritimes south to Florida and west to Louisiana and Minnesota, preferring to grow in wet meadows, moist woods, among thickets, and at the edge of moist prairies. The root is most powerful as a violent emetic and purgative when fresh. Most knowledgable collectors harvest the root in the autumn and dry and store it for up to a year prior to use. The dried root is milder in action.

HEALTH BENEFITS

The root of blackroot contains essential oils, tannic acids, mannite (a saccharine), senegin (a triterpenoid saponin), and leptandrin (a bitter glucoside). These potentially toxic constituents account for the root's being used medicinally as a cathartic, emetic, hepatic, laxative, and tonic.

BLACK SPRUCE
PICEA MARIANA

Family: Pinaceae
Parts Used: whole plant
Season: summer, fall
Region: North America

"Flat fir, sharp spruce." I am not sure where I learned this little mnemonic device, but it has always helped me when trying to tell whether I am looking at a fir or a spruce tree. Fir trees tend to have flat needles; spruce needles have

The quintessential scraggly spire of a black spruce.

square cross sections and sharp tips. Sometimes the tips are so sharp that they can pierce your skin and draw blood. Considering their naturally occurring defenses, it is a little surprising that spruce trees hold a spiritual place of honor for many American Indian communities. They are often revered as sentinels of the north and the standing guardians and protectors of the land and people. The Hopi have a Spruce Clan, and their katsina dancers are often adorned with spruce boughs. The Cherokee hold an annual spruce dance. According to a Pima story, the father and mother of the Pima people survived an earth-destroying flood by floating in a ball of spruce pitch.

USES

Spruce trees, especially black spruce, are significant and useful trees. The inner bark of black spruce is applied to areas of the skin that are inflamed and infected. The inner bark is also made into an infusion to treat kidney stones, stomach problems, and rheumatism. The resin from the trunk of black spruce is very gum-like; it is used as a poultice to heal sores, wounds, burns, and rashes. The resin can be also chewed to help with digestion and for toothaches. A decoction of the leaves, especially the young highly colored tips, can be gargled for sore throats, taken to treat diarrhea, and applied to the skin for sores and dryness. The dark cones can also be made into a decoction and taken to treat diarrhea. The young twigs can be made into a decoction and taken for coughs.

The roots of black spruce are split and made into twine and used to sew baskets as well as the ribs, gunnels, and sides of canoes. The root material is also used for sewing snowshoes. Black spruce wood makes good canoe paddles and is used for constructing drying and smoking racks, fish traps, and snowshoes. Black spruce pitch has a special quality; it has traditionally been used for making the black pitch that is used for gluing and sealing the seams of

Canoe maker Marcel Labelle paddles a birch bark canoe with seams sealed by black spruce pitch.

birch bark canoes. The pitch is boiled in order to soften it. It is then mixed with charcoal. Deer fur is sometimes added; it helps prevent the mixture from becoming too brittle when dried.

The young male catkins can be eaten raw or cooked. The newly collected inner bark, harvested in spring, is dried and pounded into a meal for thickening soups or made into cakes. The young cones can also be cooked and eaten. The young needles and even the bark are made into a beverage; the resin makes a decent chewing gum.

The dark purple, almost black cones of black spruce.

IDENTIFICATION AND HARVEST

Black spruce trees grow up to 75 feet. At higher elevations, near tree line, black spruce will grow like a large shrub; at lower elevations, this evergreen conifer takes on a scraggly, conical to spire-like shape. The branches are short and drooping. The bark is dark gray, making the inner portion of the tree appear very dark from a distance. The short, stout needles are four-angled, stiff, and sharp, with a waxy pale blue-green surface. The twigs are yellow-brown and hairy-looking. The young cones bloom during the spring; they take on a green color at first, then appear orange and finally dark purple. They will grow 1 to 1.5 inches long and will remain on the tree for several years. The small seeds ripen in October and November.

Look for black spruce across the northern latitudes of North America, from Alaska, eastward to Nova Scotia and New England, and in the states of Minnesota, Wisconsin, and Michigan. The tree does not do well in the shade and prefers wet sandy and loamy soils. It will also be found around bogs and swampy peatlands. When gathering pitch from black spruce, it is best to first scrape a small portion of the bark from the trunk. Gather the pitch that oozes in the wound that has been created. The pitch should then be boiled in water. The pure pitch will rise to the surface, where it can be skimmed off.

HEALTH BENEFITS

Black spruce's essential oil, which accounts not only for its scent but for some of its pharmacological effects, contains fifteen separate constituents with anti-inflammatory and antioxidant properties, including santene, tricyclene, myrcene, and terpinolene. Twenty other compounds (seven lignans, five neolignans, four flavonoids, three phenolic acids, and trans-resveratrol) have also been identified in this conifer.

BLACK WALNUT
JUGLANS NIGRA

Family: Juglandaceae
Parts Used: leaves, nuts, bark, wood, sap
Season: spring, summer, fall
Region: North America

Whenever I think of walnut trees, my mind immediately makes a temporal-dimensional journey to when I was a little boy. We were enjoying a family gathering in a sunlit corner of a wide canyon. At one edge ran a stream that my cousins and my brother and I were drawn to. On the other side, next to a canyon wall, stood a huge black walnut tree. Many of my family members escaped the sun by gathering in the shade cast by the tree, and we set up our tables and food there. I recall one of my aunts pointing out the large ripe walnuts and encouraging us boys to climb the thick limbs of the tree in order to collect some of the nuts. Later, we shared the company of our family while cracking open and eating the sweet meats of the nuts. It is not only my family who think so highly of walnut trees. American Indians across the continent value walnut trees for the durable and beautiful wood they offer as well as for their medicinal and other utilitarian uses. We also like having picnics in their shade.

In the Southwest, the Apache valued the black walnut tree so much that they named clans after it: the White Mountain Apache have the Chiltneyadnaye Clan, and the Pinal Coyotero Apache, the Chisnedinadinave Clan. Black walnut trees also play a role in Apache cultural narrative. They have a story about how Coyote used a walnut tree to trick a group of soldiers. One day Coyote was resting in the shade of a black walnut tree. He had just stolen a horse from a man. The saddlebags on the horse were filled with money. Coyote heard the soldiers chasing after him and came up with a plan. He placed a few coins in the branches of the tree. When the soldiers arrived to arrest

The pinnate leaves and green fruit of black walnut.

him, Coyote convinced them to wait and hear him out. He told them that if they let him go, he would tell them how to make a lot of money. The soldiers did not believe Coyote until he told them to shake the branches of the walnut tree. When they did this, the coins that Coyote had hidden in the tree fell down. Coyote then suggested that if the soldiers were to give him their pack mules as well as the packs on the mules, he would trade them the "money-growing" tree. The soldiers agreed, and Coyote made off with the stolen horse and its saddlebags, still laden with money, and a line of pack mules. The soldiers were stuck with a walnut tree.

USES

Walnut trees are used for more than tricking soldiers. Across the country, native peoples rely on the wood of walnut as a building material. And the nuts of black walnut are eaten by many eastern tribes; the Iroquois, for example, boil the nut meats and use the resulting brew as a refreshing beverage. The nut meats of black walnut are also an important ingredient in corn soups and several other traditional native dishes.

As a medicinal, the Cherokee chew the bark of black walnut to relieve toothaches and also make tea from the bark as an infusion that can be taken as a laxative. The Comanche make a poultice from the leaves and the crushed hulls of the nuts; applied topically, the poultice rids a person of ringworm. A similar compound is used to treat athlete's foot and hemorrhoids, and as an insecticide. The sap of black walnut can be used externally as an anti-inflammatory, and the leaves can be placed around the living space to rid it of fleas. The Iroquois mix the crushed nut meats with bear grease and apply it topically to repel mosquitoes. A decoction of the bark is taken as an emetic and to get rid of bile. The bark of charred twigs and old bark from the trunk is mixed with water and used as a remedy for snakebites. Many tribes use the bark of black walnut to make brown and black dyes.

In the Southwest the Apache, Yavapai, and other tribes collect and store the nuts of another walnut, Arizona walnut, for winter food. The nut meats are often mixed with mesquite flour and made into a sauce that other foods are cooked in. The Yavapai also mix the pounded nut meats with the sweet syrup of mescal agave and use it as a nutritious beverage. The bark of Arizona walnut can be made into an infusion and taken to treat rheumatism and arthritis as well as a wash for aches and pains. In the Southwest tribal peoples used the bark as well as the nut shells of Arizona walnut to make a black dye.

IDENTIFICATION AND HARVEST

Black walnut is a deciduous tree that can grow up to 90 feet tall. The central trunks of black walnut can be as much as 3 feet in diameter. In some cases, black walnut can grow up to 158 feet with an 8-foot-thick trunk. The branches are normally wide-spreading, forming a large crown. The thick brown and gray-black bark is deeply furrowed, often in a diamond pattern. Black walnut leaves grow 2 to 3 feet long; they are pinnate, with up to 20 leaflets. The leaflets are normally rounded at the base, pointed at the tip. The leaves are dark green and hairy below. Elongated panicle-like flowers, each containing 17 to 50 stamens, appear in May and June. The spherical fruits often appear in groups of two or three. Most parts of black walnut trees emit a pungent and spicy odor.

Arizona walnut (*Juglans major*) is similar but generally smaller, sometimes even growing as a shrub. Its branches are more widely spaced than those of black walnut, and its furrowed bark is not as thick. New growth is reddish brown. The pinnate leaves of Arizona walnut grow up to about 12 inches; they are ovate to lanceolate and serrate.

At least six varieties of walnuts are native to North America—all are valued by native peoples. In the East, the two native species are black walnut and butternut. Black walnut occurs from South Dakota to New York, south to Georgia and Florida, and west to Texas. It is found mostly in riparian areas, often growing

Juglans major has a large crown, with widely spaced branches.

alongside poplars, cherry, beech, and hickory trees; toward its eastern range, it will be found in floodplains growing with elms, hackberry, ash, red oak, and boxelders. Arizona walnut grows in riparian zones and along streams in Arizona, New Mexico, Oklahoma, Texas, and Utah; it is often found in deep canyons and ravines, growing alongside cottonwoods, sycamores, and ponderosa pines.

The nuts normally fall and are ready for harvesting beginning in midsummer and into the fall. They can be eaten fresh or stored in a cool dry area away from direct sunlight for several months. If the sap is collected for medicinal use, it should be used immediately; otherwise, it will dry, crack, and transform into fine dust. The bark and shells can be stored away from the sun for up to a month but will begin to lose their potency after that.

HEALTH BENEFITS

Walnuts help to reduce blood pressure, blood glucose, and lipids and are hepatoprotective and antidiabetic. Walnut leaves are important sources of several pharmacologically active compounds, among them phenolic acids, ascorbic acid, and several flavonoids, including epicatechin and aesculetin. The flavonoids offer anti-inflammatory, anticarcinogenic, antitumor, antioxidant, and antimutagenic properties. Walnut shells contain the naphthoquinones juglone and plumbagin. Naphthoquinones are antibacterials, anti-inflammatories, and antivirals; they are anthelmintic, antifungal, and even used to treat acne.

BLUEBERRY
VACCINIUM SPP.

Family: Ericaceae
Parts Used: fruit
Season: summer
Region: North America

It was a warm, humid July day high up on a shoulder of the Mad River Valley in Vermont. I was helping to facilitate a whole thinking retreat at the Refuge at Knoll Farm, a certified organic family farm whose current stewards raise Icelandic sheep and blueberries. They asked if our group would like to collect some of the blueberries that were then ripening on a quarter-acre of a gently rising hillside. I grabbed a bucket, found a row of bushes, and began picking the little blue-gray balls of juicy deliciousness. About every five or six berries, I would pause and let one of them find its way into my smiling mouth. I felt like a ten-year-old again.

USES

Everyone's favorite use for blueberries is to simply eat them, preferably straight off the bush; they can also be canned; made into jam; crushed, filtered, and cooked down into syrup; and juiced. But blueberries are much more than a sweet and tasty food. Native peoples in North America have used blueberry to treat urinary tract infections, kidney stones, colic, fever, varicose veins, and hemorrhoids. Blueberry is also used for improving circulation and memory, and as a laxative. The Iroquois

Blueberries are truly one of nature's power foods. American Indians have recognized this and used them for a variety of ailments for centuries.

use a decoction of the berries to treat itchy skin rashes. It is also used to improve night vision.

IDENTIFICATION AND HARVEST

Wild and native blueberries are normally small upright shrubs, 0.5 to 3 feet in height, depending on the variety and growing conditions (elevation, latitude, exposure). The foliage is typically deciduous, sometimes evergreen. Leaves are ovate to lanceolate, 0.5 to 3 inches long, 0.5 to 1.5 inches wide. Groups of bell-shaped flowers, greenish white, pale pink or red, dangle from the branches. The berries of the wild, lowbush variety tend to be small and pea-sized; those of the cultivated highbush variety are larger. Both varieties have a flared crown at the end of the fruit. All blueberries begin as a small pale green fruit, then become reddish purple, and finally ripen into a dark purple berry with a glaucous bloom.

Blueberries are typically a midsummer fruit. In the Northeast and Pacific Northwest, peak harvesting should be in July. Blueberries

Wojapi is a traditional Native American dish made of stewed berries; it can stand on its own as a dessert, sweetened with honey or maple syrup, or be used as a sauce on meat, game, or vegetables.

do well in open, not overly boggy areas, where their feet are damp. During pre-Columbian times, native peoples maintained their favorite collecting areas with selective and periodic pruning and coppicing of the bushes and by clearing away competing plants. Ripened berries are relatively hardy and can withstand harvesting by hand. Just don't be too clumsy or lazy, trying to collect a clump with one pull. The berries are overripe if the inside of the fruit easily squeezes out when grabbed.

HEALTH BENEFITS

Blueberries contain various beneficial phytochemicals. Most studies have been conducted using the many commercial cultivars of highbush blueberry (*Vaccinium corymbosum*); however, the straight wild species, lowbush blueberry (*V. angustifolium*), is an even richer source of anthocyanin and other polyphenols.

BLUE COHOSH
CAULOPHYLLUM THALICTROIDES

BLUE GINSENG

Family: Berberidaceae
Parts Used: leaves, roots
Season: summer, fall
Region: eastern North America

American Indians generally believe that all things that surround us are alive but especially that when something is named, it is imparted

with life. Before European contact, virtually every plant in North America had been named. The names differed according to who was doing the naming. The many herbalists who now prescribe blue cohosh are unknowingly employing a word that has been used by native peoples for millennia: cohosh means "rough" in the Algonquian family of languages. In this case, the word refers to blue cohosh's gnarled root structure.

Blue cohosh features leaflets in clusters of three and six-petaled flowers.

The common name cohosh ("rough" in the Algonquian family of languages) is a reference to the plant's root structure.

USES

It should first be noted that blue cohosh is toxic in large doses. It should not be used during pregnancy. Large doses may cause high blood pressure, nausea, vomiting, and symptoms that mimic nicotine poisoning. Improper use of the powdered root can irritate the mucous membranes.

Blue cohosh has a long history of use by eastern American Indians as a laxative, to prevent pregnancy, and for several other medical conditions. An infusion of the roots is taken to slow down excessive menstruation and to aid the mother during labor, by easing contractions. The same infusion helps relieve menstrual cramps. The roots can also help to induce labor, stop bleeding after delivery, and support the rebuilding of uterine tissue after childbirth.

Blue cohosh, as an anti-inflammatory, is also used for skin problems. An infusion of the root is used to bring down fever and is taken as an emetic, for indigestion, and as a bath to treat rheumatism. The leaves are rubbed onto an area of skin affected by poison oak.

IDENTIFICATION AND HARVEST

Blue cohosh is a perennial, about 1.5 feet tall and twice as wide. Rising on a single stalk are large, compound, tri-lobed leaflets; the bluish green leaflets are tulip-shaped, alternate, entire at the base, serrate at the tip. The flowers have six petals that are not fused; they emerge yellow then change, turning to purple to green to brown as they mature. The fruits are deep blue berries.

Blue cohosh is found in rich, moist soils, in shade. Look for it in the hardwood forest of eastern North America, from the Canadian border south to Georgia. Blue cohosh is most potent when gathered in late summer to early fall. Be sure to leave some root material behind in order to ensure future harvests. After

The berries of blue cohosh are completely smooth, unlike blueberries, which feature a flared crown on the bottom.

collection, clean the soil from the roots and allow them to fully dry in a cool, dry area. Blue cohosh roots can be stored in a cardboard box or paper bags; plastic storage causes the roots to mold.

HEALTH BENEFITS

The roots of blue cohosh contain triterpene glycosides (anti-inflammatories) and crystalline glycosides (smooth muscle stimulants). Unfortunately, they also contain teratogens, which means that consumption of the roots can cause fetal abnormalities.

BUTTERNUT
JUGLANS CINEREA

WHITE WALNUT, OILNUT, LONG WALNUT

Family: Juglandaceae
Parts Used: fruit, roots, bark, wood, sap
Season: year-round
Region: eastern North America

Cherokee novelist Thomas King (2003) had it right: our stories are "all we are." For native peoples, stories are the source of our values, morals, and identity. Stories are also the dominant path along which we transfer and reproduce plant knowledge. A person can read about plants, or listen to an herbalist teach them about plants, but unless the knowledge is related or connected to something else, chances are it will be difficult to retrieve.

A young butternut tree; mature specimens can reach 85 feet in height.

However, if that plant knowledge is transferred through a story, it is connected and related to images and commonly understood ideas—and is easier not only to retrieve but to retain.

An indigenous story in the Southeast concerns a man who liked walnuts—a lot. People from his village would often find him sitting under his favorite walnut tree, cracking and eating walnuts. He even devised a simple tool that helped him crack walnuts more easily. Inevitably, the walnut cracker died. Not too long after, another man who was passing through the area stopped beneath the walnut tree where the walnut cracker enjoyed sitting and eating his walnuts. The new man noticed all the walnuts and began to crack and eat them. He continued to do this into the night. The people from the nearby village heard the sound of walnuts being cracked and assumed that the ghost of the walnut cracker had returned. Despite the darkness, many of the villagers went to investigate the sound. Some of them were sick; one man with an injured leg even needed to be carried. They approached the walnut tree. The visitor, thinking that he was being attacked, ran off. The people of the village were startled by the sound of his footfalls, and they too fled. They were so scared, the people who were sick felt well enough to run, and even the man with the injured leg could suddenly walk on his own. Ever since this event, it's been believed that walnuts have the power to remove sickness through their very presence.

USES

The Cherokee use an infusion of the inner bark of butternut to treat toothache, as a general cathartic, and as an antidiarrheal. The Chippewa make a decoction of the sap for a cathartic. The Iroquois use a decoction of the bark as an emetic, to treat painful irritations, and to help induce pregnancy. They also chew the bark and apply it to stop the bleeding of fresh wounds. The nut meat, mixed with bear grease, is rubbed on the skin to repel mosquitoes. Oils pressed from the nuts are used to kill and expel tapeworms.

Many tribes use butternut shells for making a brown dye and the root to make a black dye. The lumber is revered for its strength, light weight, and subtle grains. Butternut is especially appreciated as a light-colored wood with incredible grains that polishes extremely well. It is also softer than its cousin, black walnut, making it a favorite of wood carvers.

Before strangers came to North America's shores, the indigenous peoples enjoyed a vast and varied cornucopia of foods, a good portion of which consisted of nuts. Every region of the continent provided at least one kind of nut for its native inhabitants. For California tribes, acorns and other nuts were a staple part of the diet. In the east, butternuts and other walnuts were an especially important food supplement. Crushed, the nut meat of butternut is added to corn stews and cornbread dough; it can also be mixed with berries and formed into small cakes for travel food. The nut meat is also boiled and puréed into a drink.

IDENTIFICATION AND HARVEST

Butternut is a deciduous tree, 60 to 85 feet tall, and short-lived, to about 70 years. The pinnately compound leaves are large (15 to 20 inches long), with 11 to 19 leaflets. The edges of the leaves have small teeth. While the leaves of black walnut feature soft hairs only on the underside, butternut's leaves are hairy on both sides. The trunk is normally ashy gray, with broad, flat plates and deep furrows in the bark. The narrow catkins of unisexual flowers are 6 to 10 inches long. Trees flower in May and June. The greenish oblong fruit (nuts) usually form singularly, sometimes in clusters of two to five fruits. The fruits are hard, with a deeply furrowed shell enclosed by a thick, sticky-glandular husk; they typically ripen in October and November and persist on the tree even after the last leaves fall.

Look for butternut in southern Canada and the northeastern and north-central United States. It will sometimes be found in Missouri and eastward through Tennessee, North Carolina, and Virginia, rarely in Arkansas, Mississippi, Alabama, South Carolina, and Georgia;

it does not form dense stands and is often difficult to find in its range. Butternut prefers moist, sandy, well-drained soils with some modicum of clay. It grows well in sunlit areas near riverbanks among mixed woodlands. There will not be much understory at the base of butternut trees.

HEALTH BENEFITS

All *Juglans* species contain the phytochemicals juglone and plumbagin, which naphthoquinones are responsible for the yellow and brown pigments found in butternut walnuts. Naphthoquinones are antibacterials, anti-inflammatories, and antivirals; they are anthelmintic, antifungal, and even used to treat acne.

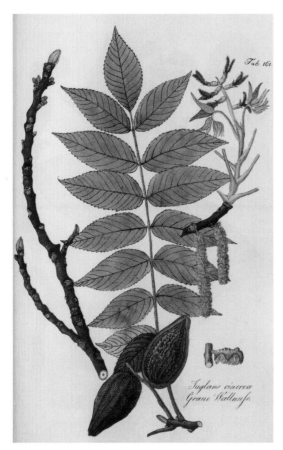

A 19th-century botanical illustration of butternut; it's easy to see why another common name for it is long walnut.

The rough, slightly oblong shell of butternut.

CALIFORNIA LAUREL
UMBELLULARIA CALIFORNICA
––––––
MYRTLEWOOD

Family: Lauraceae
Parts Used: leaves, branches, fruit, wood
Season: year-round
Region: West Coast

If you ask most people to name California's most important tree, the response will likely be sequoia or redwood, or perhaps almond trees, as its nuts are a lucrative agricultural export of the state. Ask the same question of a California Indian, however, especially one who is steeped in their culture's traditions,

A defining feature of the California laurel is its multi-trunked habit.

California laurel produces an olive-sized fruit whose inner flesh and seed are a prized food source.

and they are more likely to name the oak. Other native Californians, especially those from the Bay Area, might put forward California laurel.

I used to live with a large, multi-trunked California laurel. It was old, a century tree at a minimum, rooted on a slope near the back of the property, sharing space with a towering redwood, some oaks, and other native and introduced plants. Its trunks curled up and away from the slope; one had wrapped its way around a wooden structure. The tree offered habitat for the many songbirds in the area and shade for the few herbaceous plants that sprang up beneath it. I enjoyed the pungent scent that ever-drifted from the tree and eagerly anticipated the yearly arrivals of its small yellowish flowers and fruits. California laurels are incredible. California native peoples thought so too: these trees were sometimes essential elements in ceremony and had several other uses.

California laurel leaves are tucked underneath headbands and inside hats to relieve headaches and sudden fits. The Kashia Pomo make a poultice of the leaves to treat rheumatism and headaches. The Yuki place a single leaf into a nostril for headaches; they also bathe the head of anyone suffering from headaches with an infusion of the leaves. An infusion of California laurel can be taken to treat stomachaches, colds, sore throats, and to clear up mucus in the lungs. A decoction of laurel leaves is used to clean sores; an infusion of the leaves helps to rid a person of head lice, and a wash of the same makes an excellent bath for rheumatism. People from my tribe, the Rarámuri, often drink a tea of Mexican bay leaf (*Litsea glaucescens*), a close cousin of California laurel; it calms the nerves and, taken after meals, it ensures good digestion.

California native peoples eat the inner flesh and seed of California laurel's olive-sized fruit; the outer flesh, with high concentrations of essential oils, is often acrid-tasting and therefore avoided, unless it is first boiled, which rids it of the oils. Most often the fruits are dried until they split, revealing and loosening the large seed. The dried flesh can be eaten at this point, and the seeds are collected for roasting; when they are brown, they are eaten whole or ground or pounded into a meal. The meal is pressed into cakes, made into mush, and added to stews. Roasting removes much of the pungency of the seeds, leaving a spicy flavor reminiscent of chicory. California natives also make the meal into a chocolate-like drink. The leaves of California laurel are similar to bay leaf (*Laurus nobilis*) but are more pungent; they too can be added to soups and stews but produce a stronger flavoring.

California laurel is a hardwood yet soft enough for carving; it makes beautiful lumber products. Some California natives use the strong and flexible wood for hunting bows and drumsticks, or split it to fashion percussive clapper instruments for ceremony.

California laurel is an important part of Pomo ceremony. Branches are placed around homes for spiritual protection. To attract deer, Chumash hunters burn laurel leaves; the smoke is also said to act upon the deer like a narcotic. Other California natives burn the branches during ceremonies for spiritual cleansing.

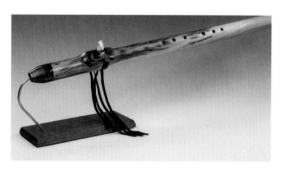

A flute made of California laurel, blackwood, and manzanita burl by artisan Tim Blueflint, a member of the Bad River Band of Lake Superior Chippewa.

IDENTIFICATION AND HARVEST

California laurel is a hardy, often multi-trunked evergreen tree that can grow up to 80 feet tall. The dome-shaped canopy

California laurel in flower.

can reach about 30 feet across at maturity. In good conditions California laurel can live to be several hundred years old. The dark green to yellow-green alternate leaves are leathery, narrow, and oblong to lance-shaped; they are 2.5 to 4 inches long, smooth-edged, and pointed at the end. Throughout the year, some leaves on the tree will turn light brown and drop to the ground. When broken or crushed, the leaves give off a pungent, peppery, menthol-like scent. The small yellow-green flowers are arranged in clusters of six to ten flowers that all spread from a common point; flowers emerge from late winter into spring. The fruits, which emerge as early as April, are olive-green and olive-shaped, about an inch long and half as wide; they mature to a brown-purple. Inside each fruit is a single large light brown seed surrounded by a light brown shell. Before trees reach maturity, the bark is thin, whitish gray, and only slightly creased; with age, the bark thickens, becoming reddish brown and peeling away in spots.

Look for California laurel along the West Coast from southern California north, along the coastal ranges, and in the Sierra Nevada; it often grows in shady hillsides among oaks, mixed evergreens, redwood forests, and even chaparral up to 5,000 feet in elevation. California laurel prefers moist but well-drained soils and does well in both full sun and shade. The evergreen leaves can be collected any time of the year, but it might be prudent to wear a mask and gloves: the heavy essential oils present in the leaves, twigs, and bark can produce headaches and skin rashes in sensitive individuals.

HEALTH BENEFITS

Twelve different kinds of flavonoids have been isolated in California laurel. Flavonoids are known to be antiviral, anti-inflammatory, antifungal, and antibacterial. Also present are quinones, alkaloids, cardenolides, tannins, and saponins, all of which demonstrate antimicrobial properties.

CALIFORNIA POPPY
ESCHSCHOLZIA CALIFORNICA
GOLDEN POPPY, CUP OF GOLD, FLAME FLOWER

Family: Papaveraceae
Parts Used: whole plant
Season: spring, summer, fall
Region: western North America

In March 2019, following one of the wettest falls and winters to date, fields and hillsides throughout the middle and lower elevations of the Golden State were set ablaze with a superbloom of yellow-orange California poppies, and people flocked to California's open spaces, parks, and national monuments in order to experience the blooms in person. But California poppy has always been a star in Chumash and other California native narrative and culture. Long before this plant was given its English or Spanish (*toroza*) common names, much less assigned its scientific binomial, it was known to the Chumash people as *qupe*, to the Miwok as *munkai*, and to the Yurok as *herk'werh 'we-chpega'r* ("cottontail rabbit's ears"). The Chumash believe that the souls of the dead have their eyes pecked out by two ravens in the afterlife and that the soul then stretches out its arms and picks a poppy to fill in each empty eye socket, whereupon it can see again. Later the flower-eyes are replaced with abalone.

USES

Although California poppy is primarily used as a medicinal, the young green leaves are foraged and eaten either boiled or steamed. Medicinally, many California native peoples make an infusion of the root and take the drink for insomnia, headaches, and toothaches, to relieve spasmodic coughs, and as an analgesic for muscle pain or injuries. The root can be made into a poultice or sliced and applied to cuts, burns, and scrapes. The juice of the root can be taken as an emetic, for stomachaches, and as a sedative. The juice is also rubbed onto a woman's breast in order to stop the secretion of milk. A decoction of the flowers is rubbed onto the scalp to kill lice.

IDENTIFICATION AND HARVEST

Individual plants of this short-lived perennial or annual are about 20 inches tall. The grayish to blue-green leaves, arranged alternately along the single central stem, are divided into feathery lobes. Large four-petaled solitary flowers, about 3 inches across, top the stems; they are usually yellow-orange to orange, sometimes pink or even red. Plants begin to

California poppy's iconic bright orange petals.

flower in February, continuing into September. Later, the plant will produce elongated, slender seedpods, 2 to 3 inches long; the mature pods split open, releasing several tiny black or dark brown seeds.

California poppy is a pioneer plant; it pops up as a volunteer in just about any xeriscaped yard, and in the wild, it brightens sandy, open, grassy, and sunny areas, vacant lots, and roadsides. Although it is native to California (the California state flower, in fact), it occurs at 2,000 to 6,000 feet elevation from southern Washington to Baja, and east to New Mexico. California poppy is best harvested during its growth stage, just before or right after the flowers have bloomed. The roots can be used fresh or dried.

HEALTH BENEFITS

The foliage and especially the roots of California poppy contain a host of alkaloids, including californidine, chelerythrine, and pavine. The plant also contains flavonoids. These constituents give California poppy antifungal, analgesic, anxiolytic, and sedative properties. Ethanol extracts of the plant have been successfully used in the treatment of heroin addiction and withdrawal.

CATTAIL
TYPHA SPP.

Family: Typhaceae
Parts Used: whole plant
Season: year-round
Region: North America

Four 13- to 15-year-old girls knelt before their families at the edge of the large ceremonial circle. At the center of the circle, a large bonfire lit the cool night in the White Mountains of eastern Arizona. I watched an elderly Apache man approach one of the girls and say something to her. She rose and carefully placed her hands on his back for a minute. The girl then returned to her kneeling position. Four Gaan (Mountain Spirit) dancers moved around the circle and among the spectators, accompanied by the rhythm of singers and drummers. The young initiates remained in their buckskin regalia at the circle's edge, their hair and foreheads dusted with a mixture of cattail pollen and cornmeal. Each young girl taking part in the

All parts of the cattail serve functional purposes, and practically all (including the heads) are edible.

Apache girls' puberty ceremony was becoming a Changing Woman; each had been spiritually invested with the pollen of a plant that many wildcrafters consider primarily a food plant.

USES

The various cattails—broadleaf (*Typha latifolia*), narrowleaf (*T. angustifolia*), southern (*T. domingensis*)—have been put to a diversity of uses by American Indians for millennia. Material culture excavated at the ruins in and around Chaco Canyon in northwestern New Mexico have included the remains of cattail pollen, shoots, and roots, and cattail has been identified in the bundles of ancestral Puebloan healers. The long durable leaves are woven into mats and baskets or lashed together (with long cattail stems) for roof thatching. In times past, some California native peoples used the downy parts of the flower head for padding in baby beds.

Almost all parts of the plant are edible. The Hopi of northern Arizona mix the heads with tallow and chew it like gum. In southern Arizona, the Pima eat the young stalks and bake the pollen into biscuits or mix it into a gruel. At the bottom of the Grand Canyon, the Havasupai eat the heads like they would corn. Over at San Felipe Pueblo, cattail shoots are ground and mixed with cornmeal. The Paiute take the seedheads and strike them with a firebrand; this separates the seeds from the down, which makes the tasty seeds easier to eat. The Paiute also grind the dried roots; the resulting flour is made into a mush or cakes. The rhizomes can also be chewed fresh, but be ready to engage in lots of spitting of the stringy pulp.

As a medicinal, cattail is used to treat kidney and urinary problems and as a dressing for wounds, boils, and burns. Different tribes apply the crushed roots to burns, wounds, and skin infections. An infusion of the dried roots and the whitish lower portion of the leaves is taken for stomach issues. The Delaware take a decoction of the roots for kidney stones; the Washoe eat the flowering heads to treat diarrhea.

Cattail down has been used to cushion baby beds.

IDENTIFICATION AND HARVEST

Look for corn dogs on a stick. Most of the plant is composed of long, slender, green stalks topped with a brownish, fluffy, corn dog–shaped flowering head. Rising from the flowering head is the dark brown male spike that holds staminate flowers. Several basal leaves surround the central stem; they are thin with parallel veins running their long, narrow length. The leaves are 0.5 to 1.5 inches wide with narrowleaf cattail, and a bit wider for southern and broadleaf cattail. In the best conditions, these colonial, rhizomatous perennials can reach up to 10 feet tall.

Cattails will be found in or near slow-moving and still waters—marshes, ponds, lakes, wet depressions—throughout North America. Late spring to early summer is the best time to collect the edible male spike of the plant. Collect the leaves, stems, seeds, fluff, and pollen from spring to early fall, when the plant head pops open; if the stems are destined to be eaten, they should be collected early in the season, before they become too fibrous. Roots can be collected year-round.

HEALTH BENEFITS

Cattails contain more calcium, iron, potassium, and carotenoids than rice or potatoes; they also contain flavonoids and alkaloids, which offer antimicrobial activity. The medicinal uses of cattail are mostly related to skin disorders, burns, and wounds: its numerous water-soluble polymers (rhamnose, xylose, arabinose) play a role in helping cells in the skin's epidermal layer regenerate and heal.

CEDAR
THUJA PLICATA, CUPRESSUS NOOTKATENSIS

Family: Cupressaceae
Parts Used: whole plant
Season: spring, summer, fall
Region: North America

Native peoples in the Pacific Northwest tell a story about a good man who gave unceasingly to his community. He was a skillful hunter and fisherman, always there to help anyone who needed it. The man eventually passed away, but the Creator, so impressed with the life this man had led, decided that a great useful tree would grow from the man's burial site. And that is where the first red cedar grew.

Another story concerns three beautiful sisters, out by the shore, drying and smoking that day's salmon catch. Trickster Raven came upon the women. Always hungry and up for some fresh salmon, Raven decided to try to separate the women from the fish. He approached them and asked if they were afraid of him. They said no. He asked if they were afraid of bears. Again they said no. He asked them if they were afraid of wolves and other scary creatures. Each time, the answer was no. Finally, Raven asked if they were afraid of owls. This time the answer was yes. Raven quickly left them and hid in a nearby wood, where he

The stately trunk of an ancient cedar.

Kwakiutl cedar mask dancers, circa 1914.

began to make loud owl sounds. The three sisters, frightened, stopped their work and ran up the side of a mountain to escape. They grew tired and eventually stopped. At that moment, they transformed into yellow cedars. This is why cedars, among the most graceful-looking and beautiful of trees, are found growing on the sides of tall hills and mountains.

Transformation is a recurring theme in American Indian narrative, worldview, art, and ceremony. It is particularly evident in stories from the Coast Salish, Kwakiutl, Haida, Tlingit, and many other Pacific Northwest

cultures. I feel the power and significance of transformation whenever I experience cedar mask dancers, who—once they don the mask of Sun, Raven, or Chief of the Sea—assume the character and spirit of the entity. The masks are always carved from different portions of the cedar tree.

USES

Ethnographers and anthropologists often designate Pacific Northwest peoples as being part of "salmon culture." It seems a satisfactory identifier, since one of the staple foods of Pacific Northwest cultures is salmon. I wonder, however, if an alternative title for these cultures might be "cedar people." Cedar predominates in Pacific Northwest material culture. Wood from the trees is made into house planks, beams, posts, roofing, canoes, and finely designed and carved bentwood boxes. The bark is transformed into rain capes, woven hats, fishing gear, and skirts; the inner bark is used as punk material for keeping fire embers alive and woven into mats and baskets. Roots and branches are made into baskets and cordage. All parts of the trees are transformed into important ceremonial items. The inner bark and leaves of cedar are sources for several medicinals.

A decoction of the scale-like leaves of western red cedar (*Thuja plicata*) is taken for coughs, colds, and diarrhea, and applied to the body for general aches and pains. A decoction of the young tips of the boughs is also taken for colds and coughs. An infusion of the leaves is applied topically to treat rheumatism and used as a hair wash for dandruff. The same infusion is taken for coughs, colds, stomach pains, sore muscles, weak heart, and as a purifying agent during sweat baths. A poultice of the young tips is applied to the stomach to help get rid of stomachaches and applied to the chest for bronchitis. The bark is chewed to induce menstruation and applied to wounds and cuts to stop bleeding. The buds are chewed to help strengthen the lungs.

The boughs of yellow cedar (*Cupressus nootkatensis*) are used in sweat baths to help relieve arthritis and rheumatism. A poultice of the leaves is applied to sores and wounds and a decoction of the leaves is taken as a general tonic.

IDENTIFICATION AND HARVEST

Western red cedars are tall evergreen conifers, growing to over 200 feet. Their crowns are conical, often becoming irregular with several leaders. Branches arch toward the ground, sometimes pendent, fanning out.

Foliage spray and seed cones of western red cedar.

The wide trunks of mature trees are buttressed at the base. The fibrous bark is reddish brown (gray with maturity), about an inch thick, flat, and separated by ridges. The small individual leaves are scaled and sharp, arranged in pairs opposite each other on fanned-out leaflets. The leaves are glossy green on top, paler below; they release a spicy piney scent when crushed. The small brown seed cones are ellipsoid, holding about ten winged seeds per cone.

Yellow cedar is also evergreen, but it is not as tall as red cedar, topping out at about 145 feet. The pendulous branches hold dark green scaled foliage that is flat, fanned out, and drooping—a sad-looking silhouette overall. Bark is gray-brown and scaled on younger trees, becoming loose and separating from the trunk in long strips as trees age. Cones are rounded and scaled; each scale ends with a pointed triangular bract. The wood has a yellow tint (hence the common name) and is powerfully pungent. As an instrument maker, I have used yellow cedar as the soundboard for classical guitars. Whenever I work with this wood, its strong spicy scent fills the air of my workshop and even moves into adjoining parts of the house.

Western red cedar is found in the Coast and Cascade Ranges, from southern Alaska and into northwestern California, at elevations up to 4,500 feet. Some populations grow in the Rocky Mountains from British Columbia and into Montana. When growing on the edge of islands and coastal shorelines, its long arching branches can be seen gracefully sweeping the surface of the waters. The tallest and finest western red cedars grow inland, where they

The scaly, rounded cones of yellow cedar.

can set their roots deep into moist soils, often forming thick groves.

Yellow cedar does well in colder, wet ecosystems in the coastal ranges of Alaska and British Columbia south into northern Oregon, where it is often found on steep hillsides, from near sea level up to 7,000 feet. It too prefers deep, moist soils but can do well where the soils are covering bedrock.

When gathering cedar for cordage, it is best to collect the branches and bark in spring, when the tree is running with sap. Parts of the tree can be collected any time of year for medicine, but the biologically active constituents are most available in the summer, when the cones have emerged. If you've harvested the bark and leaves, be prepared for their strong scent to fill your car and home.

HEALTH BENEFITS

The fresh leaves, roots, and bark of cedar contain several essential oils, including fenchone and pinene. These oils are astringent, anti-inflammatory, antibacterial, antiviral, hepatoprotective, antifungal, anthelmintic, antioxidant, and abortifacient.

CHILI PEPPER
CAPSICUM SPP.

Family: Solanaceae
Parts Used: fruit
Season: summer, fall
Region: North America

In North America, particularly in the Southwest, people are sensitive about chili peppers. Some get defensive about their peppers, feeling that theirs are the best example of hot pepperiness. Others are frightened at the prospect of coming into any kind of contact with the dreaded fruit. Most can tell some kind of story about eating peppers, either intentionally or not. Mention roasted green chili to any native of the Southwest, whether indigenous, Hispano, or Anglo, and plan for reactions ranging from delight to disdain and a litany of anecdotes from the funny and absurd to the soulful and spiritual. Some American Indians become

downright mystical about the amazing food and medicine.

I was talking with a friend, an indigenous farmer from Bolivia who now raises foods for Tesuque Pueblo in New Mexico. When I asked whether he was growing chili, his eyebrows shot up: I had gotten his attention. His normally serious demeanor shifted, and he gave me a knowing smirk. "Enrique," he said in his richly accented English, "I have something for you that you will love." Later, when we returned to his office near the Tesuque Pueblo civil buildings, he handed me a small glass bottle of orange-red liquid, indicating that it was a precious gift and cautioning me to be careful with the contents. I unscrewed the lid to take a sniff. My sinuses and eyes were simultaneously assaulted and embraced by

Jalapeño chili pepper plant.

the liquid pungency of chili oil. My friend began reciting its many medicinal uses, from a cold and flu remedy to an antibiotic. But I knew exactly what I was going to do with this ambrosia once I got it into my kitchen. The oil is all gone now, but while it lasted, it acted as a preventative medicine while enhancing the flavors of my home-cooked pinto beans, omelets, and tortilla soup.

Chili was introduced to our continent when Spanish colonists began moving north from the Tehuacan Valley of Mexico during the 16th century. The plants originated further south, in places like Bolivia, the highlands of Peru, and other tropical areas of Central and South America. The only variety of chili that existed north of the tropics before its mass migration was the chiltepin (aka chili pequin), a tiny red pepper that in maturity grows erect on small bushes. The fruits are about the size of a pinky fingernail. My grandmother used to grow these tiny balls of fire in our fields. She would grind one or two of them up to add to a pot of beans or a stew. That was enough to change a bland pot of beans into a mouth-stimulating dish. The burning sensation experienced when chilis are consumed results from their combination of oleoresin and capsaicin, an alkaloid found only in chili peppers. One has to be careful when handling chiltepins: an errant touch to one's lips, eyes, or nose can result in irritation for about an hour or so, thanks to the capsaicin found to varying degrees in all chilis.

Today these mildly irritating to powerful peppers are grown throughout South, Central, and North America by the hundreds of acres, in home gardens, and in patio pots. Chili consumption reaches deep levels of regional and cultural identity. Someone considered "local" to the Southwest, for instance, must be able to claim that they have at least tried to eat chili. Among some circles the ability to eat chilis is a sign of manliness. In the Sierra Madres of Chihuahua, Mexico, my people are wary of those that do not like or refuse to eat chili. People there feel that only evil sorcerers don't eat chili.

USES

I grow several kinds of chili peppers in my garden, for eating. Depending on the variety, they can be stuffed and baked, roasted, pickled, boiled, and dried. One of my favorite chili dishes is to simply roast the large and milder green chilis over coals until they are nearly black. I then shake them up inside a paper bag; this helps the skin peel away more easily from the fruit. I peel the peppers and eat them fresh.

Roasting over an open flame is a favorite preparation for many varieties of large, sweet chilis.

They can also be saved for adding to scrambled eggs, posole, and many other dishes. Some varieties, such as the hotter jalapeños, are chopped and served fresh. Other varieties are best when picked after they have turned red and dried on the plant. They can then be stored and ground into a powder to be added to soups and stews or as a condiment.

American Indians use chili peppers for their preventive and therapeutic properties. Chilis are eaten to prevent rheumatism, arthritis, stiff joints, bronchitis, stomachaches, chest colds, and influenza. A poultice of the plant and fruits is applied externally to treat rheumatism, fevers, headaches, chest colds, and gangrene. Chili consumption is considered good as a general tonic and for lifting a person's spirits.

IDENTIFICATION AND HARVEST

Chili peppers are grown throughout North America but thrive in hotter drier regions. Plants produce single off-white to yellowish flowers. The fruits are green at first maturing to yellow, orange, or red, depending on the variety. Cayenne pepper (*Capsicum annuum*) is a large woody herb or small shrub with lance-shaped alternate leaves. It is perennial, but to many gardeners north of the Mexico–U.S. border, it functions as a sensitive, frost-tender annual. I have the cultivar 'Del Diablo' growing in my backyard in northern California; now a small tree, it somehow manages to remain alive through the hard frosts that sometimes occur in our region. The peppers can be harvested when green or left on the plant to turn color and mature.

Mature peppers should be collected before fall rains.

HEALTH BENEFITS

Chili peppers are high in vitamins C and A and loaded with antioxidants, such as beta carotene and lutein. The principal active phytochemical, the alkaloid capsaicin, is what makes chilis a topical irritant; however, capsaicin is also analgesic, diuretic, anti-inflammatory, and antimicrobial.

CHOKECHERRY
PRUNUS VIRGINIANA

Family: Rosaceae
Parts Used: fruit, roots, bark, wood, sap
Season: spring, summer, fall
Region: North America

The land and its plant and animal inhabitants are an important source of American Indian morals and values. It can even be said that the land embodies our sense of right and wrong. These morals and values are culturally reproduced and transferred through our oral literature, our stories, and chokecherries figure in many of these. In a Northern Paiute story, Coyote is trying to get back at Bear for the murder of his son, who had interrupted Bear while he was raiding an ant nest. Coyote knew that Bear had a favorite place for eating chokecherries, and the day he learned that Bear had lashed out and killed his son, an angry Coyote went straight to that haunt. Sure enough, there

The deep purple berries of a chokecherry.

was Bear, filled with chokecherries and now napping in the shade. Coyote called out, "Is that my Aunt lying there?" Bear responded, "No, it is just me, Bear." Coyote kept talking, telling Bear, untruthfully, that he knew of a place where there were lots more sweet chokecherries. Bear was interested and demanded that Coyote tell him how to get to this special place. Coyote made up a story, saying that the chokecherries were just past a spot he knew to be a tangle of thick bushes. Bear ambled off to find the chokecherries. Coyote swiftly moved along another path, intending to reach the spot with the entangling bushes before Bear. When Bear arrived, Coyote was able to ensnare and kill him.

In a Pit River Tribe story, Fox, who is trying to get rid of Coyote, tricks him by laying out chokecherries along a path to lead him away. In an indigenous Pacific Northwest tale, Coyote fixes Magpie's broken wing with a piece of chokecherry. Afterward, Coyote learns of a large sucking monster that is wiping out one of his favorite foods, salmon. In the process of killing the monster, Coyote saves all his animal friends and creates the landscape of the Pacific Northwest as it looks today.

USES

Besides playing a role in many American Indian tales, chokecherries offer much in the way of food, medicine, and tools. At their ripest, chokecherries are very tart, even a little bitter. They become sweeter through drying or cooking. Still, they remain a staple traditional food for many American Indians, especially tribes from the Great Plains. Normally, the fruits are pounded and shaped into small cakes or balls, then sun-dried; they can then be stored, eaten as a tart snack, or included in pemmican. The dried fruits are also added to soups and stews. Fresh fruits, with sweeteners added, are made into jams, jellies, preserves, sauces, and syrups.

Tribes throughout North America employ the fruits and bark, which have astringent, anti-inflammatory, and antibacterial properties, for various medicinal uses. Both have a beneficial impact on the respiratory system. Northern Plains tribes drink the juice from the berries to treat sore throats, to stop post-partum hemorrhage, and for diarrhea. The Cheyenne sun-dry the young unripened fruits, pound them into a meal, and take it for diarrhea. The fruit can also be boiled and eaten to treat bloody bowel movements.

An infusion of the bark helps to relieve stomach problems, diarrhea, rheumatism, and dysentery. An infusion of the inner bark helps to relieve the pains associated with early labor. An infusion of the bark and inner bark is taken for coughs, colds, laryngitis, fevers, and to loosen phlegm. The root bark is also a good astringent when mixed with water and applied to open sores and skin ulcers. A poultice of the roots helps stop the bleeding of open wounds.

The flexible, sturdy wood of chokecherry is used by native peoples for making tent stakes and poles and utilitarian tools that require hard but flexible wood; it can be made into cooking utensils, tongs for grabbing hot coals during ceremonies, and carvings. The thick limbs of larger trees are ideal for carving bows; the smaller straight branches, for arrow shafts and for constructing back rests. The Lakota attach bunches of chokecherry branches to their Sun Dance poles. The sap of chokecherry makes a long-lasting adhesive.

IDENTIFICATION AND HARVEST

Three varieties of chokecherry occur in North America. The type, var. *virginiana*, produces deep red and sometimes white fruits. Var. *demissa* (western chokecherry, capulin) has dark red fruits. Var. *melanocarpa* produces deep purple to black fruit. All are deciduous large shrubs or small trees, growing to about 20 feet and often forming woody thickets of several individuals. There is no central trunk but rather many slender stems. The alternate leaves are oval to broadly elliptic, 1 to 4 inches long, and dark green, shiny on top, pale below; they can be finely serrate. The leaves will yellow in the fall. Chokecherry bark is gray and sometimes red-brown in younger trees, darkening to a dark brown-red to black with age; the bark develops horizontal grooves as it matures. The aromatic flowers are produced on new growth before the leaves emerge, usually from April to July, in cylindrical racemes up to 6 inches long; they are white and five-petaled flowers, nearly an inch across. A couple of months later, greenish drupes appear, eventually ripening to half-inch red (sometimes purple and black) fruits.

Var. *virginiana* is found primarily east of the Continental Divide. Var. *demissa* and var. *melanocarpa* are western varieties, the latter found in pockets of the Sierra Nevada foothills of California, Nevada, the eastern side of the Rockies down to the Black Range of southern New Mexico. Chokecherry grows in various ecosystems. It does well in the northern Great Plains and on higher hillsides east of the Mississippi River. It prefers locations where there is direct sunlight and moderate precipitation and often occurs in abandoned disturbed areas; in Colorado and New Mexico, I have found chokecherry growing on the edge of old mining sites at 8,000 and 10,000 feet. The wood is best harvested from late winter to early spring. Collect the berries from late summer to early fall.

Chokecherry blossoms are highly aromatic.

HEALTH BENEFITS

Chokecherry berries are full of carotenoids, phenolic acids (chlorogenic, caffeic, sinapic), and anthocyanins (cyanidin, delphinidin, petunidin), the pigments that color the berries. The woody parts of chokecherry contain healthful tannins.

CORN
ZEA MAYS

MAIZE, SUNU

Family: Poaceae
Parts Used: fruit, silk, cob, husk
Season: summer, fall
Region: North America

Corn is mother. Everyone in attendance for the first corn roast near the Hopi Second Mesa in northern Arizona is reminded of this. It is early morning. The sandy dirt is dug away, revealing the metal cover that has been placed over a deep roasting pit. When the cover is lifted steam shoots from the pit, releasing the sweet, musky scent of ears of corn that have been roasting below ground all night. Young men kneel down and then climb into the deep pit, pulling up and tossing ears of hot corn from inside. The elder offers a prayer while presenting a large ear of steaming corn, first to the east and then to the other cardinal directions. Now we can all eat.

Except for some coastal tribes from the Pacific Northwest and others farther north

Corn plays a central role in American Indian food ways and culture.

where it cannot grow, corn plays a central role in the food ways and cultures of Indians throughout the Americas, even those with nomadic lifestyles. Corn is the only traditional American Indian food plant that needs humans, planting its seeds, in order to survive. According to many tribal stories, corn has always been with us; some of us even believe that we are children of corn. But if humans were to suddenly disappear, corn, as we know it, would last perhaps another year. This is because humans created corn: according to paleoethnobotanists, corn was first hybridized

Corn has been a part of the indigenous diet for so long that American Indians have bred it into hundreds of varieties, for differing culinary uses.

about 9,000 years ago, from teosinte (*Zea luxurians*), a wild grass relative. Some think that it was somehow also crossed with eastern gamagrass (*Tripsacum dactyloides*) and possibly other relatives, such as *Z. perennis* or *Z. diploperennis*. Archeobotanical evidence suggests this crossing and selection occurred somewhere in southern Mexico, perhaps near the Rio Balsas. The practice of growing and eating corn spread from southern Mexico throughout the Americas. When Europeans first made contact with Taíno people in the Greater Antilles, they also made first contact with corn, which the Taíno called *mahis* (hence maize, another English common name for corn). My guess is that indigenous women were responsible for the creation of the spectacular new food. Native women have always been responsible for feeding the community, and to this day, it is they who are revitalizing ancestral food ways, trading heirloom seeds, and fighting for food sovereignty rights.

Corn is indeed central to American Indian beliefs, identity, culture, and foods. It is more than a food. It is also a medicine, used in crafts, and in construction. In addition, we feel that we are directly related to it. It is often a significant part of ceremony and even traditional arts. My people, the Rarámuri, believe we emerged into this world from ears of corn after a huge cleansing deluge. The Hopi believe they were asked by the Creator to choose from certain ears of corn after they emerged into this, the Fourth World; they also maintain spiritual figures known as corn maidens. The Abenaki believe that First Mother brought corn to the people. The Menominee believe that the

A Hopi yellow corn maiden katsina doll, the most ubiquitous of all the female katinsa dolls, by Benjamin Kabinto; her presence is a prayer for corn.

people received the gift of corn in a dream. To the Navajo, corn is one of the four first sacred plants—each represents a cardinal direction, and corn represents the north.

The Cheyenne have a legend about how corn came to the people. It starts with a time when the people were hungry. They could not find buffalo or other game to hunt on the plains. Even edible plants were scarce. Two young men, both wearing buffalo robes over their yellow-painted bodies, entered into a cave to visit a spring. Sitting next to the spring was an old woman cooking, tending two pots; there was buffalo meat in one pot, and in the other was a yellow food the young men had never seen before. The woman fed the two young men with the buffalo and contents from the second pot: corn. She assured them that after they left, the buffalo would return. She also told them how to plant the new food and sent them away with some cobs. Shortly after the young men returned to their village, thousands of buffalo began to spring forth from the cave. For several seasons, the people knew an abundance of both buffalo and corn, which the young men had taught them how to grow. One hunting season, when the corn was almost ready to harvest, the people left to follow a very large herd of buffalo. They returned to their village to discover that all their corn had been stolen. There was not one cob or kernel left. They tried to find the thieves but were not successful. Without any cobs of kernels, the Cheyenne could no longer grow their corn. From that moment, they would get corn only through trade.

USES

American Indians cook and eat corn in many ways—too many to enumerate here, in fact. Native peoples roast corn, separate the kernels (fruit) from the cob to roast chicos, grind roasted corn into atole and pinole, add roasted corn to soups and stews, and make roasted and ground corn into a thin bread (*piki* to the Hopi). Some people ferment corn into beer and turn corn flour into tortillas, tamales, cornbread, and mush. The Navajo make blue corn flour into small dumplings, the size of large marbles. When I go backpacking or on day hikes, I like to bring a pouch of corn that has been soaked in salt and slowly parched into a centuries-old version of corn nuts. On Sunday mornings, I still like to whip up a batch of blue cornmeal pancakes served with maple syrup.

Corn smut (*Ustilago maydis*) is an edible gray tar-like fungus that grows among the kernels on ripening ears of corn. Many people refer to it by its Nahuatl name, *huitlacoche*. It forms when moisture from rain seeps down in between the protective husks on ears of corn. In many parts of the Southwest and Central America, huitlacoche is considered a delicacy. In some farmers' markets in Mexico, a small bucket of the fungus can fetch three dollars. It is cleaned, cooked with spices, garlic, and onion, and then added to quesadillas, tacos, and stuffed chicken.

Corn smut, an edible fungus, is considered a delicacy in many parts of the Southwest and Central America.

Corn silk is collected during the growing season from the tops of the emerging ears. The silk is made into a decoction and then taken for bladder and urinary tract infections, to treat kidney stones, and even to stop bedwetting. The same decoction is taken as a general tonic and used to treat a weak heart, diabetes, and high blood pressure.

Dried corn cobs can be soaked in water and rubbed on the skin to treat poison ivy. The Tewa use blue cornmeal mixed with water for bodily pains and sore throats. Cornmeal is also used to treat diaper rash, rheumatism, and eczema.

Many tribal peoples roll a smoking mixture into a dried corn husk for ceremonial use. The husks are also used for wrapping tamales and making children's dolls. Braided husks are used to make ceremonial masks, sleeping mats, baskets, and sometimes footwear. Every aboveground part of the plant is used in ceremony in one way or another.

IDENTIFICATION AND HARVEST

Corn is really a large grass: it's in the same family as the grass on your neighbor's lawn, bamboo, and wild rice and other grains. Corn is a true annual: it must be planted by humans every year. Like other grasses, corn has a hollow stem with alternate leaves. Corn flowers, or tassels, form at the top of the plant, appearing like a small mop. The newly forming ears are also arranged alternately on the sides of the central stem. Since *Zea mays* was first hybridized by indigenous Americans thousands of years ago, hundreds of cultivars have been selected for their distinctive shape, size, color, yield, and ability to grow in specific conditions; however, these selections may be said to fall into six major categories: dent, flint, pod, sweet, flour, and popcorn.

I enjoy growing one of my people's 41 varieties of corn every year. Many tribes who rely on corn as a staple will conduct up to three separate plantings. An early planting often takes place in late March, a second planting about a month later, and then a final planting at the beginning of summer. When to plant is often determined by phases of the sun and moon. Southwestern tribes (e.g., Hopi, Tewa, Zuni, and Navajo) plant several seeds in each deep hole on the Colorado Plateau; when the stalks grow to maturity, they will support each other in the windy conditions of this region. Most American Indian corn is produced for grinding into flour, so the cobs are left on the stalks into the fall; however, the Cherokee and other tribes collect and eat green corn in July. To tell if corn is ready to be harvested, feel the end of the cob. If it feels pointy, it needs to continue growing. If it feels rounded or blunt, the cob is ready for picking, roasting, and eating.

HEALTH BENEFITS

Corn silk contains flavonoids, alkaloids, phenols, steroids, glycosides, terpenoids, and tannins. These phytochemicals help reduce inflammation and are antioxidant, antilithiatic, uricosuric, and diuretic.

Tending corn planted in a traditional field on an arid plateau in Monument Valley, Utah, circa 1952.

COTA

THELESPERMA SPP.

Family: Asteraceae
Parts Used: whole plant
Season: spring, summer, fall
Region: Southwest

My two Hopi friends make their home on the flat area that separates Second and Third Mesa in northern Arizona, an arid area where they are nonetheless able to coax about an acre of heirloom blue corn up from the sandy soils. On a recent visit, I noticed the plant I knew as cota growing wild near their house. When I mentioned it, they said that they call it Hopi tea (I was to later learn that the Navajo call it Navajo tea, and the Zuni, Zuni tea, and so on). The reason for the riot of common names is that this plant grows throughout the Colorado Plateau and other regions of the Southwest. It is known to nearly all peoples indigenous to the area.

USES

Before the introduction of coffee and black tea to the Southwest, cota was a popular drink. Cota has a distinctive sweet piney flavor. An

Cota thrives in rocky soil at high elevations.

The flower of *Thelesperma filifolium*.

infusion of the entire plant is drunk with honey or sugar as a relaxing beverage. It is also taken to soothe the nerves, as a diuretic, and to reduce fever in children and babies. Some drink cota to clean the kidneys and urinary tract. Cota also makes a reddish brown dye.

IDENTIFICATION AND HARVEST

Cota (aka greenthread) does indeed have thin, thread-like leaves scattered along a single stem. It is a single-stemmed (occasionally multi-stemmed) herb, 1 to 2 feet in height, depending on the species and growing conditions. The stems of *Thelesperma filifolium* are topped with a small golden yellow aster-like flower with a reddish brown center. The flower of *T. megapotamicum* is actually several small florets with purple-tinged phyllaries; *T. longipes* is similar except that the leaves tend to grow in the lower half of the stem and the phyllaries tend to be dark green. The center florets and phyllaries of *T. subnudum* are often surrounded by five to eight larger (half-inch long) ray florets, slightly fringed at their

margin. Cota's taproot helps it to survive in arid environments, and bloom time is long, from April to September. This drought-resistant plant grows in open fields, rocky slopes, and open forest meadows at 3,000 to 9,000 feet. It can be found in the Four Corners region and from Nevada north into Wyoming and western Montana. It is best to collect cota during its long flowering period, from late spring to early fall. Harvest the entire plant.

HEALTH BENEFITS

Little is known about the phytochemicals in *Thelesperma* species. Cota does contain some phenols and is mildly antiseptic.

COTTON
GOSSYPIUM SPP.

Family: Malvaceae
Parts Used: leaves, stems, fruit, roots
Season: summer, fall
Region: Southeast, Southwest

Cotton is most widely known for the soft fibers of its seed coats, from which textiles are made; however, the cotton plant also serves American Indians as a source of both food and medicine. Pima cotton (*Gossypium barbadense*) was probably domesticated by native peoples in the upper third of South America. The North American native cotton, *G. hirsutum*, was probably first hybridized somewhere in southern Mexico about 5,000 years ago; it was an important trade item among early Mesoamerican cultures, such as the Maya, Olmec, and Mixtec.

USES

Cotton has long been used as both a textile and fiber for making cordage by the Pima, Tohono O'odham, Hopi, and other Pueblo peoples. The long fibers are used to weave ceremonial clothing, sashes, mantas, and string for prayer sticks. The fibers are also used for making large ceremonial baskets at Santa Clara Pueblo and for oil lamp wicks among the Navajo. The Pima and Tohono O'odham toast the high-protein seeds to be eaten as a snack or to be ground into flour for making cakes and bread.

Cotton is perhaps best known for its woolly white seed pods, but these are preceded on the plant by whitish to yellow flowers with striking purple centers.

Gossypium barbadense is one of two species native to the western hemisphere.

Different parts of the cotton plant are used medicinally as well. Some Southeast peoples take a decoction of cotton roots to ease childbirth. The Tewa chew the seeds into a poultice to be applied to the heads of children to cure hair loss. The young shoots of the plant are made into an infusion to treat asthma; the crushed seeds are taken to relieve sore bones. The leaves can be boiled and added to a bath to ease convulsions. The leaves are also used to bring down fevers and to treat coughs, colds, and skin diseases and ulcers. Stems are crushed into a powder to treat snake and scorpion bites. Cotton root can be used as a hemostat and to treat nausea, fevers, headaches, diarrhea, dysentery, and nerve pain.

IDENTIFICATION AND HARVEST

Cotton is an annual or perennial herbaceous shrub that grows to about 5 feet. The plant normally sprouts several branches. The entire plant is covered with small black oil glands. The three-lobed leaves are arranged spirally on the branches; the lower leaves can sometimes have five lobes. The lobes are ovate, sometimes triangular, each ending with a sharp apex. The solitary five-petaled flowers emerge during the summer; they are whitish to yellow with a purplish center. The fruit, a capsule about the size of a walnut, contains three to five cells, each containing many half-inch-long, black to dark brown ovoid seeds. Each seed is covered with woolly hairs.

Gossypium hirsutum and *G. barbadense* are both native to the western hemisphere. *Gossypium hirsutum* grows in the Southwest; *G. barbadense* is found in Florida and other southern states where frosts are rare. Both are domesticated and part of large agribusiness. The plant generally stops flowering in September; this is the time to look for the large capsules beginning to open.

HEALTH BENEFITS

The complex phytochemical gossypol is present throughout the plant, with the highest concentration occurring in the seeds. Pharmacologically, gossypol is antitumor, spermicidal, antimalarial, antiparasitic, antiviral, and antiamoebic.

CRANBERRY
VACCINIUM SPP.

Family: Ericaceae
Parts Used: leaves, stems, fruit
Season: summer, fall
Region: North America

Ask an indigenous person, perhaps someone from the Northeast, to choose plants that epitomize persistence, and there's a good chance that cranberries will top the list. For centuries prior to European contact, cranberries were part of pemmican recipes, and they keep finding their way into our cultures and kitchens. They were there at the fabled first Thanksgiving. They still take a star turn on our fall and winter holiday dinner tables and are with us year-round as part of granola mixes and energy bars.

USES

First and foremost, cranberries are a food plant. Cree, Penobscot, Passamaquoddy, and other Algonquian-speaking peoples eat the berries fresh, include them in stews, and dry them for winter storage and to be mashed up, along with wild game meats, to sweeten and flavor pemmican cakes. The juice of the berries is used as a dye for blankets, rugs, and clothing. An infusion of the twigs and leaves is used to treat inflamed lungs as well as to relieve nausea.

IDENTIFICATION AND HARVEST

The cranberry is a low-growing woody perennial. Its small oval evergreen leaves are arranged alternately on fine vine-like shoots that take root at intervals as they crawl along the ground, often forming a dense mat. The leaves are dark green above, pale below. One to four flower buds form on short, upright shoots at the tips of stems from May to July. The small, pale pink flowers look like shooting stars, with petals that are bent backward; the shape of the flower inspired the colonists' original name for the plant: craneberry. The Algonquin word for it is *sasemín* ("sour"). Bright red berries ripen from late summer to October. Species include *Vaccinium macrocarpon* and *V. oxycoccos*.

Cranberry plants are found in scrub-shrub swamps and bogs and marshy shores of ponds and lakes of North America. The plants do well in warm, sunny climates; they do not tolerate shade. Old stands will be found in peaty bogs where there is a conifer overstory. Cranberries

Cranberries have long been recognized by American Indians as a highly nutritious food source.

are best harvested by hand from late summer to early fall, though the evolutionary quirk that makes this berry float has led to today's commercial method of harvesting by flooding bogs and skimming the berries off the surface of the water.

HEALTH BENEFITS

The berries are nutrient-rich, high in vitamin C, fiber, and potassium. The plant and berries are also rich sources of polyphenols that display antibacterial, antiviral, antimutagenic, anticarcinogenic, antitumor, antiangiogenic, anti-inflammatory, and antioxidant properties. Cranberries also contain flavonols, anthocyanins, benzoic acid, ursolic acid, phenolic acids, and terpenes—the phytochemicals behind their demonstrated ability to fight urinary tract and other infections.

CURRANT

RIBES SPP.

Family: Grossulariaceae
Parts Used: leaves, stems, fruit, roots, bark
Season: spring, summer, fall
Region: North America

When I lived in Colorado, one of my favorite times of year was late summer to early fall, when berries ripened in the higher elevations. Armed with any kind of bucket and buoyed by thoughts of the sweet/tart syrup that would soon be poured over pancakes, we would venture up to about 10,000 feet, to our favorite stands of black currant bushes. For American Indians, an annual berry-picking foray is a necessary communal and subsistence event. It is a time for families, relatives, and clans to tell stories and be social. It is also a time to gather a nutritious food and medicinal source for the coming winter. Some cultures mark the change of season by the ripening of certain berries. Black currant marks the end of summer and the beginning of fall in the Mountain West.

The nearly black berries of *Ribes americanum*.

Several species of currants are native to North America. Some argue that currants should be separated taxonomically from gooseberries. Nevertheless, both remain in the genus *Ribes*. The difference is that gooseberries have spiny branches and solitary flowers, whereas currants are spineless and their flowers are clustered. Note: although many *Ribes* species are commonly known as currants, these are not to be confused with the dried, raisin-like Zante currants found in European scones, cakes, and puddings; that berry is actually *Vitis vinifera*, the common wine grape.

Currants are a delightful food for American Indians. The Lakota eat black currants (*Ribes americanum*) fresh off the bush, or dry them for later use in soups, stews, cakes, and as an important ingredient in pemmican; they eat Missouri gooseberries (*R. missouriense*) in a similar fashion. Tribes throughout the continent eat the berries of golden currants (*R. aureum*) and other species. The berries are also made into jams, preserves, and syrups. But currants afford other uses as well.

The fuchsia-like flowers of *Ribes lobbii*.

The roots are scraped and rubbed onto open sores. The roots of black currants are made into a decoction and taken by Blackfoot women for uterine troubles. The same decoction can be taken for kidney ailments. An infusion of the roots and bark is taken for vomiting, for kidney troubles, and to expel intestinal worms. The juice from the berries is used topically to treat cuts and wounds, or drunk as a laxative. A decoction of the leaves and stems of currants is taken as a diuretic. The inner bark of woody parts is made into a decoction and rubbed onto sores and taken to treat paralysis. The Blackfoot make a poultice of the root bark of golden currants and use it to treat swellings; they use the root of northern black currant (*Ribes hudsonianum*) to treat kidney disease and for menstrual cramps. The Cree use the fruit of skunk currant (*R. glandulosum*) to enhance fertility; a decoction of its stems is taken to reduce postpartum blood clots.

As with currants, different varieties of gooseberry are eaten by tribes across North America. In addition, the inner bark of spreading gooseberry (*Ribes divaricatum*) is chewed and swallowed for sore throats. A decoction of the root and bark is made into an eye wash. An infusion of the roots of gummy gooseberry (*R. lobbii*) is taken to treat diarrhea. A poultice of the root ash is rubbed onto boils and applied to mouth sores. The Ramah Navajo use a decoction of the leaves of orange gooseberry (*R. pinetorum*) as a ceremonial emetic.

The Cheyenne and Navajo make arrow shafts from the woody stems of black currants and gooseberries. Gooseberry roots are pounded with juniper roots and rose roots and woven into cordage.

IDENTIFICATION AND HARVEST

Currants are erect, multi-branched shrubs, 4 to 8 feet tall. Leaves are simple, alternate, lobed, dark green above, slightly hairy below, and about 3 inches wide. In May black currant bushes sprout clusters of small tubular white flowers with five petals; golden currant has yellow flowers. Smooth greenish berries are produced in drooping clusters in late summer. In August and September the small fruits of black currant turn from red to purple to nearly black; golden currant berries ripen from green to yellow to red and then finally purple.

Gooseberry bushes tend to be wide and spread out, growing only to about 5 feet tall. The thick branches bear sharp spines, either singly or in groups of two or three. The large leaves are rounded with three to five lobes. Dangling from the groups of leaves are solitary greenish to pink or red fuchsia-like flowers. Gooseberry bushes normally produce small hairy greenish fruits; some varieties produce fruits that are yellowish, white, red, and dark purple to black.

North American currants often form open thickets at the edges of shaded woods in the higher elevations of evergreen and mixed deciduous forests. Some varieties of golden currant (e.g., *Ribes aureum* var. *gracillimum*) are endemic to California; the yellow-flowered and black-berried clove currant (*R. aureum* var. *villosum*) occurs from western Texas north to

Montana and eastward to New York and Vermont. Different varieties of gooseberries are found throughout North America and as far north as the Arctic Circle. Take care when harvesting gooseberries, as the spines are sharp and stout.

HEALTH BENEFITS

Currants and gooseberries contain high concentrations of polyphenolic compounds (anthocyanins, flavonols, vitamin C), phenolic compounds (protocatechuic, quercetin, catechin), and organic acids (citric, tartaric, malic). Concentrations of anthocyanins and flavonols tend to be higher in black currants; golden gooseberries have more phenolic acids. All currants are high in prodelphinidin, a tannin with demonstrated anti-inflammatory, antiviral, and antiproliferative effects.

Deer tongue in full purple bloom.

DEER TONGUE
TRILISA ODORATISSIMA
———
VANILLA LEAF

Family: Asteraceae
Part Used: leaves
Season: summer, fall
Region: Southeast

I am not certain who introduced me to the idea of adding deer tongue to my smoking mixture. Perhaps it was my old friend Bill Buffalo (not to be confused with Buffalo Bill, although my friend's personality was just as big). What is certain is that ancestral blends of smoking mixtures (aka kinnikinnick) can sometimes be a bit harsh, and somewhere along the way, I learned to sweeten them with this leaf. The addition of deer tongue removes some of the harshness and makes the scent very pleasant.

USES

Besides adding a vanilla-like scent to smoking mixtures, the dry leaves are used as a purifying smudge by some southeastern tribes. An infusion of the leaves is taken as a general tonic, febrifuge, and for coughs.

Deer tongue is an herbaceous perennial, about 3 feet tall in ideal conditions, characterized by its basal cluster of 6- to 8-inch-long tongue-shaped leaves. A central flower stalk rises from the middle of leaves; each inflorescence bears about 30 purple flowers, which bloom in late summer. Deer tongue leaves release a pleasant scent when broken up or crushed. Some people think it is reminiscent of vanilla; others are reminded of freshly mown hay. Look for deer tongue in the low pinelands and pine barrens of the Southeast, from North Carolina to Florida and west to Missouri. It often grows in semi-shade; in moist soils, it can take full sun. The leaves can be collected any time during its growth period, but are best after the flower stalk has bloomed.

The leaf of deer tongue is a valued addition to American Indian smoking mixes.

The leaves contain coumarin, which accounts for the pleasant scent that deer tongue imparts to smoking mixtures. Coumarin is a stimulant, demulcent, and febrifuge, as well as being diaphoretic.

DOGFENNEL
EUPATORIUM CAPILLIFOLIUM

MAYWEED, SUMMER CEDAR, CYPRESSWEED

Family: Asteraceae
Parts Used: whole plant
Season: spring, summer
Region: Northeast, Southeast

Dogfennel looks and even kind of smells like the common fennel of culinary fame, but it is actually from a different plant family entirely. This is why everyone should be wary of common names: they can be dangerously misleading. Unlike common fennel, which can be beneficial to the liver, dogfennel contains toxic constituents that can damage the liver if used improperly. Nevertheless, many native herbalists from the Southeast rank dogfennel among their favorite useful plants.

USES

The whole plant can be rubbed onto exposed skin as an insect repellent. In addition, the crushed leaves and stems of the plant are

The feathery fronds of dogfennel may resemble those of common fennel, but they are not even in the same family.

applied topically to relieve insect and reptile bites and to treat fungal infections. An infusion of dogfennel is taken to treat reproductive system ailments, as an aphrodisiac, to reduce fevers, to treat epilepsy, and for sore throats. Dogfennel is sometimes used as a culinary herb, but given its aforementioned potential for toxicity, this is not recommended!

IDENTIFICATION AND HARVEST

Dogfennel is normally an upright billowy perennial shrub, 4 to 10 feet tall, with several stems emerging from a central root ball. The shiny thread-like leaves are deeply dissected. In the fall, white star-like blooms grace the ends of the stems. When crushed, the leaves give off a strong odor, similar to camphor. Dogfennel grows in open areas in full sun, such as at the borders of woods, clearings, disturbed fields, and roadsides. It is native from Massachusetts south to Florida and west to Texas.

Dogfennel blooming in early fall.

HEALTH BENEFITS

Dogfennel contains pyrrolizidine alkaloids, which can be toxic to the liver and to pregnant and breastfeeding women; however, these phytochemicals are also antimicrobial, anticancer, and anti-inflammatory. The plant also contains cyclic monoterpenes (camphene, limonene, cymene, ocimene), which are among the essential oils that give it its fennel-like scent. Monoterpenes also demonstrate antifungal, expectorant, bactericidal, antiviral, and antiseptic properties.

ECHINACEA
ECHINACEA SPP.

Family: Asteraceae
Parts Used: whole plant
Season: spring, summer, fall
Region: North America

Echinacea does not grow in the landscape of my people; therefore, although I was raised with lots of plant knowledge, I was unfamiliar with it. Only when I began to augment my knowledge base with other American Indian nations' teachings did I learn how widely used it is, medicinally and as an item of trade, among American Indians from coast to coast. It is thought to be a powerful spirit medicine, which is why many native peoples include a piece of the root in the medicine pouches they wear around the neck or on the waist. The Navajo consider echinacea to be a life medicine, one of the 40-some ceremonial plants used at certain times and for specific ritual healings; at least six of the life medicines must be used

Echinacea flowers are highly attractive to pollinators.

A beaded medicine pouch from the Great Plains.

On the Great Plains, a poultice or wash of the root of *Echinacea angustifolia* (narrowleaf purple coneflower, elk root, scurvy root) is applied to swollen areas, deep cuts and wounds, burns and sores, and painful muscles. The same poultice and wash is used to clean infected wounds. The roots are chewed in sweat lodge ceremonies to aid in the purification process. Both roots and seeds are chewed to relieve toothaches, sore throats, coughs, and colds. An infusion of the dried and powdered leaves and roots is used for sore necks and muscles. The roots and leaves are part of an antidote for snakebite as well as for venomous insect stings. The roots of *E. pallida* (pale purple coneflower) are made into a decoction used to reduce fevers and to treat arthritis, rheumatism, burns, boils, and colds. The root is also chewed to relieve toothache and colds and to increase the production of saliva during Sun Dance ceremonies. A decoction of the whole plant is used as an antidote to rattlesnake bites and to treat head lice, burns, and sore eyes. *Echinacea purpurea* (eastern purple coneflower) is used similarly to *E. angustifolia* and *E. pallida*. The roots of all species are made into an infusion and taken to strengthen the immune system and to relieve flu and cold symptoms.

IDENTIFICATION AND HARVEST

Echinacea is a perennial herb that grows up to 2 feet tall. It has a deep woody taproot and simple leaves arranged alternately on a single (usually) rough-hairy, unbranched

at the same time for peak effect. Surprisingly, echinacea does not grow wild in any part of the Navajo homeland—they had to trade for it. Just one more reflection, then, of how significant this plant has been to American Indians.

stem. The leaves of *Echinacea angustifolia* leaves are lance-shaped, 2 to 12 inches long and about 1.5 inches wide. The leaves of *E. purpurea* and *E. pallida* tend to be ovate to lanceolate and serrate, up to 10 inches long and about an inch wide; the leaves of *E. pallida* will be quasi-heart-shaped at the base. Echinacea flowers typically resemble purple to pale pink sunflowers with 3-inch-wide heads. Some endemic echinaceas have reddish or yellow flowers. Plants bloom in June and July.

Echinacea is generally found on the prairies and plains east of the Rocky Mountains from Montana to Texas and east to the coast. It is best to harvest echinacea leaves when the flowers are in bloom. The roots should be collected in the fall after the flower heads have dried up. The roots are deep and fibrous, often requiring a shovel or even a pickaxe to dig up. The leaves and roots can be used fresh but remain very effective when dried.

HEALTH BENEFITS

Echinacea species contain pyrrolizidine alkaloids, alkamides, caffeic acid, and polysaccharides. The alkaloids can be toxic to the liver and to pregnant and breastfeeding women (see Joe Pye weed), but they are also antioxidant, antibacterial, and antiviral, among other pharmacological effects. It is the alkamides, caffeic acid, and polysaccharides that play a direct role in echinacea's immune-boosting and anti-inflammatory activities; echinacea is particularly effective in preventing upper respiratory tract infections.

ELDERBERRY
SAMBUCUS SPP.

Family: Adoxaceae
Parts Used: whole plant
Season: year-round
Region: North America

It took me some time, but I was finally holding a straight branch of elderberry of the right size and length. I had already carved the bowl of my first smoking pipe, but I needed a stem. That's where the piece of elderberry came in: an elder had told me that elderberry makes the best pipe stem because the wood is light yet strong. Also, and most important, its center is composed of a soft pith that is easily removed, leaving a hollow stem through which to suck the smoke. The elder showed me how to heat the end of a long piece of wire and to use that to carefully burn out the pith along the entire length of the stem. He smiled at my disbelief that it was actually working. After about an hour of poking and reheating, I had hollowed out the pith and was ready to finish constructing my pipe.

Elderberry is considered by people around the world as one of the natural world's best medicines. In North America, indigenous

The author's own hand-carved elderberry smoking pipe.

peoples take advantage of elderberry's many medicinal gifts as well as the sweet juices and syrups it provides. Elderberry also provides material for important craftwork.

USES

The berries of elderberry are a favorite food of American Indians. Crushed, strained, and boiled into a syrup or made into jam, they are added as a sweetener to tasty cakes and many other recipes; the syrup is also said to repel many kinds of illnesses. Native peoples use different parts and preparations of elderberry medicinally, and the wood, stems, branches, and twigs are used for musical instruments and for ceremonial purposes; the stems of elderberry are especially useful when hollowed out to make smoking pipes, flutes, blowgun darts, and arrow shafts. Elderberry leaves are made into a decoction and used in steam baths to sweat out colds and headaches. An infusion of the leaves is used as a purgative and to treat skin infections, swellings, burns, and other skin problems. An infusion of the flowers is used to bring down fevers and to treat colds. An infusion of the berries is taken for rheumatism, as a diuretic, for urinary tract infections, and as a general tonic. An infusion of the bark is taken for diarrhea and as an external analgesic.

Unripe elderberries are white and waxy-looking; as they mature further, they turn a deep, dark blue to purple-black.

An elderberry bush bursting with fruits in various stages of ripeness.

Do *not* harvest any part of *Sambucus racemosa*, as the stems, roots, foliage, and berries are toxic to various degrees.

IDENTIFICATION AND HARVEST

Elderberry is a deciduous shrub, to 9 feet tall and nearly as wide. Plants often grow close together, in stands. The leaf structure of elderberry is reminiscent of walnut or ash trees: the opposite, pinnately compound leaves are 6 to 12 inches long; leaflets are serrate. In June and July, creamy flat-topped inflorescences, about 4 inches across, dot the bush. The bunches of berries, each about 6 millimeters in diameter, appear in August and September; berries are green at first, becoming white-waxy, then powder blue, and finally blue to purple-black at full maturity. Three types of elderberries are native to North America: black elderberry (*Sambucus nigra*), blue elderberry (*S. nigra* ssp. *cerulea*), and red elderberry (*S. racemosa*).

Look for elderberry in open places in riparian areas, along and above the banks of streams and rivers. It does well among (but not directly below) trees below 9,000 feet in elevation, in sun to semi-shade. Red elderberry does have some pharmacological benefits but should be avoided entirely; its fruits, in particular, are very toxic.

HEALTH BENEFITS

Elderberry contains several flavonoids (catechin, quercetin, kaempferol) that act as anti-inflammatories and antibacterials. Both berries and flowers are stocked with immune-boosting anthocyanins, which are antioxidant, anticancer, antimicrobial, and antiviral.

FIDDLEHEAD FERN

MATTEUCCIA SPP., *OSMUNDASTRUM* SPP., *PTERIDIUM* SPP.

Family: various
Parts Used: fronds
Season: spring, summer
Region: Northeast

I still vividly recall the first time I ate fiddle-heads—what a pleasant sensation it was for my mouth and for my taste buds! The soft curling heads are slightly chewy but easily break down with a couple of bites. A gentle bitterness mixed with complex earthy tones settles onto the edges of the tongue. It is similar to when one eats morels, with a hint of asparagus as well. It should come as no surprise that the taste of fiddleheads is similar to mushrooms and asparagus. Fiddleheads are actually the young sprouts of fern fronds. Ferns, like mushrooms, reproduce by means of spores and spreading rhizomes. They do not have seeds or flowers. Fiddleheads are more than a wonderful addition to the dinner table. They offer much in the way of essential nutrients and medicinal value.

USES

Fiddleheads are primarily an early spring food source, a tasty, fresh green addition to a meal after a long winter, which in the past too frequently consisted of eating dried meats and reconstituted and canned plant foods. To prepare, first rinse the fiddleheads: small bits of detritus can get caught up in the furls. Next, boil them for seven to ten minutes; this step is important, as the bitters and toxins often found in older ferns (e.g., ptaquiloside, or PTA) can be leached from the sprouts. Finally, sauté them in butter or olive oil. I like to cook them with a little white wine and shallots, but many native peoples simply boil fiddleheads and add them to spring meals.

Fiddleheads have another use, however. When the plants have grown into mature ferns, Abenaki herbalist Judy Dow likes to harvest the fronds in midsummer to make baskets.

The tender young fronds of *Matteuccia struthiopteris* are toxic if not fully cooked.

Yurok hat, partially woven with mature fern fronds.

Hupa hat, partially woven with mature fern fronds.

A collection of Karuk baskets, circa 1923.

She collects the dark stems of mature ferns, allows them to dry, and then pounds each stem, to separate the dark outer husk from the softer and lighter-colored inner strands. It is these inner strands that are often dyed and incorporated into some of the intricate designs seen in the work of experienced weavers.

IDENTIFICATION AND HARVEST

Fiddleheads tend to grow in small clumps of about six plants. They like damp areas away from roads and other disturbed areas. Once out in the woods, search for the little 3- to 4-inch-tall curled fronds pushing their way up from the forest floor near springs, along creeks, and in any damp area that does not receive too much sun. Imagine the tops of small bright green violins, or fiddles, sticking up from the ground. Remember, fiddleheads are really sprouting ferns, so look for those shady and damp areas where ferns would typically grow. One way to identify fiddleheads is to look for the light brown paper-thin material that covers the outside of the furled plant; this will come off on its own or can be brushed off. There should also be a slight groove along the stem.

After the harsh Northeast winter has subsided and during the early spring warming, when the land is still moist, fiddlehead gatherers trek to their favorite fiddlehead collecting sites in search of the young curling heads, emerging through the leaf litter of the previous fall; the ideal time in New England is normally around mid-April. Be ready to collect on short notice: fiddleheads sprout and unfurl very quickly, and once unfurled they become bitter.

Fiddleheads that are already unfurling should, therefore, be avoided. Be careful to harvest only the part of the plant that has emerged above the soil. Leave the belowground part of the plant so that it can return next season. To further ensure future harvests, never pick all the plants in a stand.

HEALTH BENEFITS

Fiddleheads are a good source of antioxidants and insoluble fiber. The furled heads are also chock-full of omega-3 and omega-6 fatty acids and high in iron, as well as containing the essential minerals magnesium, potassium, selenium, and phosphorus. Note: certain species of fiddleheads, such as ostrich fern (*Matteuccia struthiopteris*) and bracken fern (*Pteridium aquilinum*), have been shown to be toxic and possibly carcinogenic if not fully cooked.

GOLDENROD
SOLIDAGO SPP.

Family: Asteraceae
Parts Used: whole plant
Season: spring, summer
Region: North America

American Indians use goldenrod as a gambling medicine, among other things. Games of chance precede the modern-day eruption of Indian casinos by thousands of years. One of the biggest criticisms that Europeans made against American Indian culture was their

Most of the more than 100 species of goldenrod native to North America have densely clustered individual yellow flower heads.

constant gaming and gambling. But American Indian games of chance must be perceived through an American Indian lens. Gambling is not only a source of entertainment and community building, it is a sacred practice that is representative of the unpredictable Trickster consciousness and those unknown and unexplainable gray areas of the cosmos. Gambling is sacred chance, an opportunity to be in contact with the living, breathing, scattered cosmos. It is understandable, then, that a native person might need and value some gambling medicine.

USES

More than 100 species of goldenrod occur in North America; the uses expressed here refer to the genus generally and to specific species when appropriate. Goldenrod root is made into a poultice and applied to burns. An infusion of the leaves is used as an antiseptic and astringent for wounds, cuts, and bleeding. The infusion is consumed to treat urinary tract infections, excessive mucus, influenza, whooping cough, kidney and bladder stones, gastritis, rheumatism and arthritis. Similarly, a tea of the flowers is used as an analgesic and astringent, to reduce fevers, and to treat diarrhea, muscle pain, and snakebites. The Zuni and others chew the flowers to treat sore throats.

The Chippewa make a poultice of *Solidago altissima* and apply it to burns, ulcers, and boils. The Cahuilla use the leaves of *S. californica* as a hair rinse and for toothaches, burns, wounds, and boils. The seeds of *S. canadensis* are eaten by the Goshute. The Iroquois use the roots and flowers as an emetic, to treat liver problems, and as a gambling and love medicine; the roots are also taken as a sedative. The Navajo keep a piece of *S. canadensis* as a good luck charm for gambling and use the roots as part of a smoking mixture. The Menominee use the dried and powered leaves of *S. flexicaulis* as a snuff for headaches and the whole plant to treat fevers. The Iroquois make an infusion of the leaves of early goldenrod (*S. juncea*) to relieve nausea; they use a decoction of the flowers as an emetic and the roots for fevers. The Cherokee use an infusion of *S. odora* to treat coughs, colds, nervousness, and fevers; the root is chewed for sore throat. The stems of *S. gigantea* are used to make baskets.

IDENTIFICATION AND HARVEST

Goldenrods are generally herbaceous perennials, usually to about a foot in height, occasionally up to 5 feet tall (e.g., *Solidago canadensis*);

some have a crawling habit. Goldenrods grow as single-stemmed plants branching toward the top of the plant. The leaves and stems are smooth with some tiny hairs on the tops of the leaves; the leaves sometimes have wide serrations toward the tips. Some species drop their lower leaves just before flowering. The flower heads contain two to 35 florets; some species have up to 60; the heads are grouped in compound inflorescences arranged in multiple racemes, panicles, corymbs, or arrays, where the florets are situated on one side. The florets resemble small daisies, with distinct ray and disc florets; they are usually yellow, but a few species (e.g., *S. bicolor*) have white centers.

Goldenrod grows in open fields, meadows, prairies, and savannas. It prefers moist soils and damp meadows, areas near waterways, and wet ditches along roads and other disturbed areas. Plants are sometimes found growing on semi-dry, open slopes in upland prairies and deciduous and evergreen forests. Goldenrod is a long-lived perennial that will continue to grow in the same place year after year. It is best to gather the leaves and flowers in the summer and dry them for later use. Gather seeds in the fall when the flowers dry up.

HEALTH BENEFITS

Solidago species contain several biologically active compounds: the flavonoids (quercetin, kaempferol) are hemostatic, styptic, antihypertensive, antimicrobial, and vasodilators; the phenolic glycosides are anti-inflammatory; the saponins are antifungal. The leaves and flowering tops are anthelmintic, antiseptic, diaphoretic, and mildly diuretic; they act as a febrifuge and stimulant. Seeds are anticoagulant, astringent, and carminative.

GROUNDNUT

APIOS AMERICANA

HEPNISS, INDIAN POTATO

Family: Fabaceae
Parts Used: fruit, roots
Season: year-round
Region: eastern North America

Look into the worldview of virtually all American Indian belief systems, and you will discover in each at least one knowledge domain connected to the idea that humans are related to the natural world. In some cultures, such as mine, the idea is reflected in the language, where adjectives used to describe human relatives are also applied to certain plants (e.g., osha); other cultures have named their clans (culturally sanctioned extensions of human kinship systems) after particular plants. The Eastern Creek tell a story about a time when a new group of people arrived in their territory. These people were friendly and were permitted to live in peace with the Creek, but the Creek could not figure out how to include the newcomers and their children into their existing clan. Soon marriages occurred between the Creek and the new people, and mixed-blood children were born, adding to the turmoil and confusion. After some of the mothers of the classless people left the area looking for a sign,

The groundnut's maroon to chocolate-brown flowers feature a distinctive hooded central petal.

the Creek created a new clan, the White Potato (aka Groundnut) Clan, as a way to create kinship with the new people.

USES

Groundnuts are primarily a tasty food source for many native peoples east of the Mississippi River. Although they are not in the same family as potatoes, they are collected, processed, and eaten in much the same manner. Groundnut tubers should not be eaten raw; they are much better when boiled or baked, with a taste similar to that of roasted sweet potato. They can be dried and stored for later use or dried and made into a flour for soups, stews, breads, and cakes. Some tribes with access to birch or maple syrup bake groundnuts in it to produce a dish reminiscent of candied yams. Some New England tribes use the tuber medicinally, for proud flesh, a severe skin condition; the tuber is boiled, made into a poultice, and then applied to the cauliflower-like growth. As a legume, the plant also produces fruit pods full of pea-like seeds that are collected and cooked like beans.

IDENTIFICATION AND HARVEST

Groundnut is a climbing perennial that forms low thickets, often climbing over other plants and trees; the vining stems can grow as long as 10 feet. The pinnate leaves are alternate,

Groundnut tubers are often processed and eaten in many of the same ways as potatoes.

ovate, 1 to 4 inches long, 0.5 to 2.5 inches wide, and often hairy. Typical pea flowers form in rounded clusters among the leaves from July to October; they are white to reddish brown with two brown-purple side wings, two brownish red lower petals, and a brown-red

A 19th-century botanical illustration of groundnut's leaves and flowers.

Groundnut fruit pods are full of edible pea-like seeds.

sickle-shaped keel. The slightly curved fruit pods are 2 to 5 inches long and about a half-inch wide; they begin green but dry in the fall, splitting open to release the dark brown edible seeds.

Groundnut is mostly found east of the Mississippi River, occasionally in the central prairie states from Texas to North Dakota. Look for it in wet meadows, near streams, ponds, and sloughs, and in the moist soils of prairie ravines and woodlands. The seeds can be collected in the fall when the pods spring open. Groundnut tubers can be harvested any time of year but are best when dug up during the fall. The tubers are distributed every 10 to 12 inches along the rhizomes; they should be scrubbed clean of dirt and the small nitrogen-fixing nodules before consumption. To ensure future harvests, be careful not to damage the roots you are leaving behind.

HEALTH BENEFITS

Groundnut tubers are 13 to 17 percent protein, a good source of complex carbohydrates, and high in soluble fiber. But again, do not consume raw groundnut tubers: the tubers contain protease inhibitors and glycosides that can cause severe stomach upset and even allergic reactions. When cooked, these compounds are leached out. Groundnuts are also a good source of proline, an amino acid that helps build collagen, which then promotes good skin and digestive health, prevents joint pain, improves the cardiovascular system, fights inflammation, and lowers blood pressure. The seeds contain as much protein and fiber as pinto beans.

HACKBERRY
CELTIS SPP.

Family: Ulmaceae
Parts Used: leaves, fruit, bark, wood
Season: summer, fall
Region: North America

In common American discourse, the word *desert* often has negative connotations—desolate, lifeless, dry. But the Apache, Tohono O'odham, Maricopa, Pima, and other native peoples thrive in the Sonoran Desert, which also happens to be one of the most biodiverse regions in the world. I've led my ethnobotany students on many hikes through it over the years; they

The bright orange fruit of *Celtis laevigata* var. *reticulata*.

decoction of the bark of common hackberry (*Celtis occidentalis*) is taken to treat sore throats and colds and to regulate menses. The berries of desert hackberry (*C. ehrenbergiana*) are very sweet; they are either eaten directly or added to other dishes as a sweetener. The berries of common and netleaf hackberry (*C. laevigata* var. *reticulata*), as well as being eaten fresh, are dried to be made into cakes or used as a flavoring for dried meats. The berries can also be made into preserves. The hard wood of hackberry is a good fuel for long-burning fires; it is also used for making fences and handles for tools. The Navajo make a red-brown dye from the leaves and branches of netleaf hackberry. The Tohono O'odham make sandals from the bark of desert hackberry.

IDENTIFICATION AND HARVEST

Desert hackberry (aka spiny hackberry) is a small (to about 10 feet) deciduous tree with spiny whitish gray branches; the half-inch spines don't discourage the many desert birds who feast on its sweet orange and yellow berries. The bark is smooth and gray, the leaves small and rounded. In spring, the tree sprouts clusters of inconspicuous white flowers; fruiting begins in summer and continues into fall.

Common hackberry is also deciduous but a much larger tree, up to 55 feet tall, with young branches forming a drooping zigzag pattern. The bark is warty and gray; the leaves are alternate, 3 to 5 inches long, sharply toothed, dark green on top, pale beneath. Small greenish yellow flowers emerge in April and May with the leaves. The edible fruits are small and round,

are always amazed by how much the desert has to offer. One of the plants I make a point of highlighting during these walks is the desert hackberry: the field schools always occur in summer, when desert hackberries are loaded with their sweet little orange berries. Other kinds of hackberries are native to North America as well; all are edible and offer medicinal and utilitarian uses.

USES

The leaves of hackberry are made into an infusion and taken to relieve indigestion. A green at first, maturing to dark red or black in September and October.

Netleaf hackberry grows more like desert hackberry, making a small tree or large shrub with smooth twigs free of spines. The bark is ridged and gray. The leaves, up to 3 inches long, have smooth margins and a slightly asymmetric base; they are rough and darker green on their upper surface. Small, nearly inconspicuous flowers emerge with the leaves; the round quarter-inch green fruits begin forming in early summer, turning reddish during late summer and into the fall.

Common hackberry is native to much of North America; it is usually found in bottomlands but also occurs on upland slopes and bluffs, limestone outcrops, and rocky hillsides. Desert hackberry normally grows well in dry rocky soils near washes and gravelly slopes; it is most often found in the desert regions of Arizona, New Mexico, and Texas, among native desert scrub and semidesert grasslands, at 1,500 to 4,000 feet elevation. Netleaf hackberry is native to the drier regions of Arizona, California, Nevada, New Mexico, Utah, Kansas, western Texas, Oregon, and Washington; look

The immature green drupes of *Celtis occidentalis*.

for it along dry rocky streambanks, washes, canyons, and limestone hills.

The useful leaves and bark should be collected during the summer and into the fall. Wait until late summer to collect the sweet berries; the fruit harvest continues through fall.

HEALTH BENEFITS

The phytochemical constituents of other *Celtis* species have demonstrated antimicrobial and antioxidant activity, but no similar studies have been conducted with common hackberry, desert hackberry, or netleaf hackberry.

HORSENETTLE
SOLANUM CAROLINENSE

———

CAROLINA HORSENETTLE, BULL NETTLE, DEVIL'S TOMATO

Family: Solanaceae
Parts Used: leaves, fruit, roots
Season: spring, summer, fall
Region: Southeast

Languages are fascinating windows into how a culture expresses the reality of the universe. In other words, language is culture and worldview. In many American Indian languages, there is no word for poison, nor are there any words for poisonous plants; if words are windows into how a culture thinks about things such as plants, then we can infer from this that to American Indians all plants—even dangerous ones—must have some kind of beneficial purpose. This is the case with horsenettle. The fruits of this plant are toxic. They have been known to cause the death of toddlers. However, the same toxic constituents, in the hands of an experienced herbalist, can heal certain ailments.

USES

The Cherokee make an infusion of the leaves to rid the body of worms and to heal ulcers. An infusion of the tiny seeds from the fruits is gargled for sore throats and taken for goiter; the same infusion can be rubbed onto skin stricken with poison ivy. A decoction of the leaves and sweet milk is used to attract and kill flies. The Cherokee string the roots of horsenettle around a baby's neck to ease the pain of teething. The fruits and roots are used to treat epilepsy, bronchitis, and convulsions.

IDENTIFICATION AND HARVEST

Horsenettle is in the nightshade family, which includes potatoes, tobacco, and tomatoes. It is an herbaceous perennial that grows up to

The spines on the stems and leaves of horsenettle deter herbivores.

Horsenettle's flowers are pale white to pale lavender and shaped like stars.

Horsenettle, a member of the nightshade family, produces toxic fruits that resemble small tomatoes.

3 feet tall. The plants sprout in the spring, producing alternate leaves that are oblong and irregularly lobed, about 4.5 inches long and 1 to 3 inches wide. The leaves are dark green on top, lighter green underneath, and hairy at the margins; the center of the leaves and the stems contain little prickly spines. Pale white (sometimes pale lavender) star-like flowers with yellow or orange centers are arranged in clusters. Small green fruits that resemble tomatoes are produced later in the summer; they turn

yellow as they mature and ripen. The fruits are just over a half-inch in diameter and are full of small flat seeds.

Horsenettle is often found in dry fields with sandy soils, grass pastures, and hay fields. It is native to the Southeast but grows from New England to Florida and west to Texas and the Dakotas.

HEALTH BENEFITS

The active constituents of horsenettle are glycoalkaloids, specifically solanine, which is more active in plants during late summer and fall. Stems and roots are less toxic than leaves, which contain less solanine than the berries. Unripe berries are more toxic than ripe ones. Alkaloids are known to demonstrate antispasmodic, antibacterial, and analgesic properties.

HORSETAIL
EQUISETUM SPP.

Family: Equisetaceae
Parts Used: whole plant
Season: year-round
Region: North America

American Indians recognize that the land and its plants, and even the people, did not simply reach their current form on their own. We all needed a little help from powerful beings and energies. One of those was the Trickster being Coyote. But to describe Coyote as merely a Trickster ignores many positive and often not-so-positive qualities that the indigenous

Horsetail features distinctive dark rings between the joints of segments.

concept of Trickster represents. One of his more admirable qualities is that Coyote is responsible for either bringing into existence many parts of our current landscape or playing a role in shaping or reshaping parts of it. That includes our plants.

Horsetail (aka scouring rush) grows throughout North America, including the Kootenai homelands of Montana. However, indigenous stories tell that the plant used to be a plain-looking simple green tube of a plant. It had no beautifully regular rings or purplish tint like it does now. That is until one day, when

Coyote, down along a rushing river looking for food, accidentally fell into the water. The swift current was pulling him downstream and underwater. Fearing for his life, he reached out toward the bank, grasping desperately at some horsetail plants. As he attempted to pull himself onto the bank, he noticed that the plant was barely keeping him from slipping back into the churning water and perhaps to his death. He spoke to the plant and made a promise: if it held him while he pulled his way back onto dry land and safety, he would use his powers to make the plant and all its relatives beautiful. The horsetail held, and Coyote was saved. Coyote kept his promise and gave horsetail the interesting look that characterizes the plant today.

USES

Horsetail contains a large percentage of silica. As a result, the plant can be used to scour pots and pans. It also makes a good substitute for fine sandpaper. Many native peoples use this plant as a fingernail file or for polishing wood carvings, hunting bows, arrows, and stone pipe bowls. The hollow tubes of some species of horsetail are large enough to fashion whistles and for weaving into mats. The rhizomes of horsetail are used as material for basket weaving and are also boiled or roasted for eating. The very young shoots are also eaten like asparagus. The plant can be made into a poultice and applied externally to stop bleeding and to treat rashes, head and muscle aches, cuts, poison ivy, and burns. An infusion of the plant is taken as a diuretic and an eye wash for sore eyes; for bladder, kidney, and urinary tract infections, gonorrhea, backaches, lumbago, colds, coughs, stomach problems, and constipation; to halt internal bleeding; to expel the afterbirth; and to treat rheumatism. A decoction of the plant is used to treat menstrual irregularities, gonorrhea, sore eyes, and itching skin sores; to relieve painful or infrequent urination; to speed up a difficult labor; and to expel the afterbirth. It is also mixed with chokecherry for colds and rubbed into the scalp to rid it of lice. An infusion of the rhizomes is taken when there is blood in the urine.

IDENTIFICATION AND HARVEST

Several *Equisetum* species are native to North America. In general, horsetail is an evergreen perennial characterized by green to dark green jointed, hollow, single stems that grow 1 to 4 feet tall. The stems are topped with a sporangia that produces the plant's spores; this rounded node resembles a small pine cone. Distinctive rings occur along the length of the stems; the area around these joints often has a purplish tint. Long segmented leaves sprout in whorls at the joints; these start dropping in late fall and continue dropping through winter. The leaves are replaced with growths that resemble pointed teeth and eventually return in the spring. Horsetail grows in groups from dark fibrous rhizomes; look for it in wet areas at the edge of marshes, streams, rivers, and ponds, where its long rhizomes can reach water but still maintain their central roots in sandy, gravelly, and clay soils. In temperate zones horsetail can be collected any time of the year but is best harvested for eating when the young

Horsetail prefers to grow in moderately damp areas.

shoots are just emerging in the early spring. Otherwise, the plant should be collected while the leaves are still vibrantly green and before they begin to droop downward.

HEALTH BENEFITS

Working in conjunction with each other, several phytochemicals in *Equisetum* species (kaempferol, quercetin, triterpenoids, caffeic and silica acids) are pharmacologically active as analgesics, anti-inflammatories, antioxidants, antimicrobials, antidiabetics, astringents, antihemorrhagics, and diuretics.

Indian hemp produces panicles of white flowers in midsummer.

INDIAN HEMP
APOCYNUM CANNABINUM

DOGBANE, HEMP DOGBANE, PRAIRIE DOGBANE, HONEYBLOOM, BLACK HEMP, LECHUGUILLA

Family: Apocynaceae
Parts Used: whole plant
Season: spring, summer, fall
Region: North America

Sometimes a plant or animal epitomizes and shapes a culture. Without the historic presence of the buffalo, the nomadic lifestyle of the Lakota and other Plains Indians would have been drastically different from what it was. In the Sonoran Desert of southern Arizona, the saguaro continues to be central to Tohono O'odham connections to their arid landscape. Pacific Northwest cultures, from Alaska through British Columbia to northern California, remain deeply reflective of their staple food, salmon; however, if not for a single plant and the fibers that it provides, this connection to salmon might be irretrievably frayed. That plant is Indian hemp.

Narratives expressed all along the coast of the Pacific Northwest concern the role that Raven played in bringing salmon to the people, but if not for Indian hemp (and Coyote), the salmon might have disappeared. Recall the story mentioned in passing at chokecherry: Coyote saved the animals, including salmon, from a dangerous sucking monster. During that adventure, Coyote had to challenge the monster to a sucking contest. Coyote told the monster that he would go first by trying to suck the monster into his mouth. Of course, Coyote failed; he could not do this. But then came the monster's turn. Before the monster tried to suck Coyote into his mouth, Coyote tied a rope, made of Indian hemp, around his waist. The monster inhaled deeply, sucking Coyote into its mouth and down into its stomach. There Coyote used the rope to rescue his animal friends.

USES

In times past, before their flows were impeded by dams or siphoned off for agriculture, West Coast rivers teemed with salmon. The most efficient way for American Indians to capture this incredible food source was by twisting and braiding the strong stem fibers of Indian hemp into rope, which was then tied into miles of nets and weirs. It was and remains a source of fibers that is carefully managed through seasonal burning and by careful pruning and coppicing, so that the following year's plants will grow straight and tall. Besides being used for fishing, the fibers are made into bow strings, baskets, and clothing.

Indian hemp is also a useful medicinal. A person's hair can be rinsed with a decoction of the root to prevent it from falling out. The same decoction is taken as a laxative, as a

Indian hemp cordage created by the Soboba Band of Luiseño Indians of southern California, circa 1917.

ceremonial emetic, and to relieve rheumatism; it can also be taken for coughs, dropsy, earaches, nervousness, colds, and to rid the body of intestinal parasites. A decoction of the leaves can be applied to a breastfeeding woman in order to induce a better flow of milk. The Navajo make a decoction of the entire plant for stomachaches. The leaves can be dried, made into a powder, and snuffed for coughs and colds. The powder is also used to dress sores, skin ulcers, and wounds. A poultice of the leaves is also used to treat hemorrhoids and applied to the eyes to treat soreness and infections. Finally, some Pacific Northwest people collect and eat the seeds of Indian hemp.

IDENTIFICATION AND HARVEST

Indian hemp is an herbaceous perennial that can grow up to 6 feet tall. The reddish stems support broad, opposite, lanceolate leaves, 1 to 6 inches long; they are dark green and smooth on the top with whitish hairs underneath. The leaves have a conspicuous white central vein. In July and August, Indian hemp produces panicles of small urn-shaped flowers with large sepals and a five-lobed white corolla. Later, slender cylindrical fruits emerge. The fruits grow up to an inch long. At maturity, the fruits split open, revealing seeds that are connected to silky tufts of hair.

Indian hemp grows at lower elevations throughout the United States and into Canada. It does well in damp locations at the edges of wooded areas and hillsides with gravelly and sandy soils. It can also be found in damp roadside ditches and near streams, ponds, and

The silky white hairs of Indian hemp's seeds.

springs. Indian hemp is collected for medicine when in flower. Wait until the fall in order to collect the stems for their fibers. Cut the stems, allow them to dry for a few days, and then split or gently pound them open to reveal the long fibers. The fibers can then be twisted into string, which is itself twisted and braided into cordage and rope. Be careful when collecting Indian hemp: the stems release a milky latex that causes skin blisters in sensitive individuals.

HEALTH BENEFITS

Indian hemp is applied by knowledgeable healers as an emetic, diaphoretic, antispasmodic, cathartic, laxative, and antidiarrheal, but care must be taken when using it: its main pharmacologically active constituents are apocynin, apocynamarin, and cymarin. Apocynin is anti-inflammatory, antibacterial, antiarthritic, and effective with atherosclerosis; however, apocynamarin and cymarin are cardiac glycosides, which can negatively affect the heart.

IRONWEED
VERNONIA SPP.

Family: Asteraceae
Parts Used: whole plant
Season: spring, summer
Region: eastern North America

My ethnobotanical colleague and friend Karen Adams once told me that a weed is really "just a useful plant with a lousy press agent." I believe ironweed fits that description. To many gardeners, ironweed is a nuisance. It sprouts up everywhere—especially in rural areas—taking over open areas and gently sloping hillsides. To American Indians, however, ironweed is a very useful medicinal. Depending on the reference, 18 to 25 species of ironweed are native to North America. I will focus on broadleaf ironweed (*Vernonia glauca*), giant ironweed (*V. gigantea*), Missouri ironweed (*V. missurica*), and New York ironweed (*V. noveboracensis*).

USES

Cherokee herbalists use an infusion of the root of broadleaf ironweed, giant ironweed, and New York ironweed to relieve post-childbirth pains. An infusion of the root is also taken as a blood tonic, for stomach ulcers, and to prevent menstruation. The Cherokee also use the infusion of the root of New York ironweed to treat dysentery. The Kiowa make a decoction of the entire plant of Missouri ironweed and use it to treat dandruff. The flowers can be used to make a purple dye and as a sweet snack. Ironweed is ingested to help pass kidney stones; other medicinal uses include reducing fever and chills and treating gout, headache, and skin rashes.

IDENTIFICATION AND HARVEST

Ironweeds are perennial plants, with hairy and often coarse stems that are 3 to 6 feet tall; the stems of giant ironweed, although not as hairy, can grow up to 10 feet. Ironweed blooms from July to September. The small purple

Ironweed, as its name suggests, is seen as a pernicious nuisance by many, but it actually has a host of medicinal applications.

composite (disc and tubular) flowers grow in groups of 20 to 40 per head; each flower has a five-lobed corolla and five stamens. The alternate leaves, 3 to 10 inches long, are lanceolate and serrate. The root systems are extensive. Ironweeds are found throughout the states east of the Mississippi River and in Missouri, growing in prairies, grasslands, open fields, overgrazed pastures, and roadsides. Giant ironweed is often a midwestern plant, although it is also found in New York. Broadleaf and New York ironweed are found in the more eastern states from New York to Georgia and west to Alabama and Pennsylvania. The plant and roots are best collected while the plant is in flower. Leaves begin to dry up and die with the first frost.

HEALTH BENEFITS

The roots of *Vernonia* species contain pharmacologically active tannins, alkaloids, and flavonoids. These components have demonstrated antimicrobial and analgesic activity, are good blood purifiers, and can help prevent atherosclerosis.

JOE PYE WEED

EUTROCHIUM MACULATUM

—————

BONESET

Family: Asteraceae
Parts Used: stems, roots
Season: spring, summer, fall
Region: North America

Look for the blooms of Joe Pye weed on impressive 2- to 6-foot-tall stems.

Humans have named plants in numerous ways. The common names for most plants can typically be traced to a cultural group or originate from a specific geographic region. Often the name is connected to a story about the plant or reflects how it is used or, perhaps, the morphology of the plant. Sometimes, however, we have no clue how a plant received its name. This is the case with Joe Pye weed. According to early 19th-century botanical texts, Joe Pye was an itinerant huckster, a traveling snake oil salesman who cured people using the plant. Similar tales have it that Joe Pye was a native healer from one of the New England tribes who managed to cure typhus using the plant that later bore his name. An Anishinaabe herbalist suggests that the plant's name is derived from Zhopai, an Abenaki medicine man who had great success treating typhus using the root of this plant. It has even been put forth that Joe Pye is a mispronunciation of eupatory, an earlier common name for this genus: in the past, "eu" was often voiced as "ju," hence jupatory—and eventually Joe Pye.

Most likely the name Joe Pye comes from Joseph Shauquethqueat, a Mohican leader who was born (probably in Connecticut) in 1722.

As indigenous surnames were often difficult to pronounce and spell, many native peoples adopted local surnames when signing documents. Py, Pye, and Pie were all familiar names during this era, and records indicate that several Mohican families adopted one or another version as a surname. Shauquethqueat's family later moved to near Stockbridge, Massachusetts, where Joseph eventually became an indigenous leader (likely one of several) who relied on the herb lore of his culture to treat fevers and other ailments using Joe Pye weed.

USES

Abenaki, Algonquin, and Cherokee healers use a decoction of the root to aid recovery after childbirth. The root is also used to treat rheumatism and as a diuretic. The Cherokee use an infusion of the root for kidney trouble and as a tonic during pregnancy. The Chippewa use a decocted wash of the root as a sedative for children; it is also used to settle stomach gas. The

Joe Pye weed's root can be used to create a variety of infusions and decoctions.

Iroquois use an infusion of the root to bring down fevers and for liver problems. Finally, the hollow stem of Joe Pye weed is used as a straw to suck water from springs.

IDENTIFICATION AND HARVEST

Joe Pye weed is an erect perennial, 2 to 6 feet tall, with a hairy and purple-spotted central stem. Fragrant red-purple flowers are arranged in flat-topped panicles of at least nine florets per head. Plants flower from July to September. The ovate, heavily veined and toothed leaves grow to about 8 inches and are arranged in whorls of four or five around the central stem. The root is composed of several tan to brown fibrous rhizomes. Joe Pye weed is found in undisturbed moist and wetland habitats, from Nova Scotia south to the mountains of North Carolina and from Nebraska northwest to British Columbia. The related *Eutrochium purpureum* var. *carolinianum* is found in the Piedmont area of North and South Carolina. The roots of Joe Pye weed grow down into the soil about 10 inches. Use a trowel for collection, being sure to leave some of the root behind for future harvests.

HEALTH BENEFITS

The roots of Joe Pye weed and other *Eutrochium* species contain several pyrrolizidine alkaloids, which are often hepatotoxic. Low-level use of these types of alkaloids is generally not immediately hazardous (the potential for developing liver disease comes

Panicles of red-purple florets are topped by fine white hairs.

The tall flower heads of Joe Pye weed often tower above others in an open field.

with long-term use of the plant), but mothers should avoid exposing their unborn or nursing children to pyrrolizidine alkaloids, or risk fatal neonatal and/or infant liver injury. *Eutrochium* species are also rich in terpenes, phytosterols, and sesquiterpene lactones, which phytochemicals show anticancer, antiparasitic, and antimicrobial activity. The aforementioned alkaloids, although potentially toxic with long-term use, also demonstrate anti-inflammatory effects.

JUNIPER
JUNIPERUS SPP.

Family: Cupressaceae
Parts Used: leaves, fruit, roots, wood
Season: year-round
Region: North America

"There's another one!" I shout. My students, who were jammed along the seats and benches of the fifteen-passenger van, were craning their necks and squinting their eyes trying to figure out what I was talking about. "Look—over by that big outcrop." Finally one of them said, "All I see are those big scraggly bushes." "That's it," I said. "One-seed juniper." The confusion stemmed from the fact that, for the previous two weeks, I had been referring to juniper as a tree in the classroom. Many of the students in this summer ethnobotany field school were from outside the Southwest. Their concept of "tree" differed greatly from that of the people who were born and raised on the Navajo reservation we were then driving across.

Juniper holds a very special place in the minds of native peoples, Hispanics, and other multigenerational residents of the Southwest community. It is tough and resilient, with many practical and sacred uses. It shows up in several American Indian origin stories and has even been used as an analogy explaining why American Indian peoples will always occupy this arid landscape.

Juniperus monosperma is one of the taller plant forms on an arid landscape.

The berries of *Juniperus monosperma*.

I was at Taos Pueblo in northern New Mexico one winter. A Pueblo elder and I were following the syncopated movements of a small group of buffalo dancers in the central plaza. After being amused by some tourists who were complaining that there were not enough colorful dancers to take photos of, my friend made this comment: "Indians are like the juniper tree. Our roots are deep and strong. When the next big wind comes across the land, we will still be standing."

Other native peoples also hold the juniper in high regard. To the Navajo, juniper is one of the five sacred medicines because it survived a cataclysm. According to Navajo storyteller Hoskie Benally, Mother Earth and Father Sky were embroiled in an argument over what parts of the Creator's realms they controlled. Mother Earth naturally insisted that everything on land was hers; Father Sky likewise insisted that everything that moves above the land and in the air was his. They were so upset, they did not speak to each other for four years. During those four years, Creation began to change. The air got stale, the rain stopped. Everything became dry and lifeless. Living creatures on and below the soil began to disappear. The plants too began to disappear. Finally everything was gone except for four plants and a bird. The plants and the bird spoke with Mother Earth and asked her to end the dispute with Father Sky so that things might return to normal. She agreed and suggested that they also speak with Father Sky about the need to make amends and end the selfishness that was destroying all Creation.

The plants decided that the bird would fly up into the sky and speak with Father Sky. The bird flew up and up, higher and higher, until the plants could no longer see it. The bird was gone for four days. After the fourth day, the plants noticed rain clouds building to the south. They heard thunder. They saw a flash of lightning. From the lightning emerged the bird, now an eagle. As the eagle flew back down toward the land, it was followed by the clouds, thunder, lightning, and finally rain. Everything began to come alive again. Everything that had been lost on the land returned. The four plants who had approached Mother Earth for help were tobacco, yucca, sagebrush, and juniper. Afterward, it was decided by the Creator, Mother Earth, and Father Sky that because of the ability of those plants and the eagle to survive, and because of their resilience, they would always be central elements in all ceremonies and would be important spiritual medicines.

USES

Indeed and still, juniper is present at nearly every kind of indigenous ceremony in the Southwest. Prior to entering into the lodge, many sweat leaders will offer the participants a cup of an infusion made from the leaves of juniper. This is said both to purify the body and to produce a better sweat. Infusions of juniper leaves are also used as a diuretic and to treat skin and urinary tract infections. The same infusion is also used by many elders to soften the pains associated with arthritis and rheumatism. Drinking the infusion as well as smudging with the leaves is often used to purify the body as well as a room or house, to keep bad thoughts at bay, and to ward off spirits and energies with ill intentions.

The wood of juniper is very aromatic and is used when a hot fire is needed. Many traditional Pueblo potters prefer juniper logs when firing their finished pots in an open kiln. In addition, juniper is a medium hard wood. Some traditional American Indian flute makers construct wonderful-sounding flutes out of the straighter-grained pieces of juniper. Ghost beads are made from the dried berries of juniper. Among traditional Navajo, for example, one might see little children wearing bracelets and necklaces of juniper berries in order to ward off evil spirits. Juniper roots are pounded

Juniperus deppeana is known for its deeply fissured bark.

with gooseberry roots and rose roots and woven into cordage.

IDENTIFICATION AND HARVEST

Of the approximately 25 *Juniperus* species native to North America, about 12 are native to the Southwest. For this entry I will focus on three of the most culturally salient in the greater Southwest: one-seed (*J. monosperma*), Utah (*J. osteosperma*), and alligator juniper (*J. deppeana*). Junipers can look like big scraggly bushes to the uninitiated, although some can grow as tall as 25 feet. They usually inhabit rocky outcrops but can be found along disturbed trails and in small stands in open areas. Juniper roots are deep, far-reaching, and allelopathic (i.e., they produce phytochemicals that stunt or terminate the growth of other plants); therefore, the areas immediately surrounding individual trees will not contain much in the way of herbaceous understory or other competitive plants. One-seed juniper and Utah juniper typically produce several trunks that emerge from a single location at the base of the tree. Alligator juniper usually sports a single trunk and can often be identified from a distance by this characteristic; upon closer inspection, look for its distinctive fissured bark, which does resemble alligator skin.

Although many people (especially Southwest locals) refer to junipers as cedars, they are not true cedars (*Cedrus* spp.), which hail from the Mediterranean region and Central Asia. Leaves of the junipers noted here are scale-like and, unlike some other North American species (e.g., *Juniperus communis*, *J. virginiana*), not sharp to the touch. Another identifier of the genus is the green to bluish round berry produced by these trees. In one-seed juniper they are nearly perfectly round, but in Utah juniper, a definite point protrudes from one side of the spherical berry. As coniferous evergreens, their leaves, wood, and berries can be collected year-round.

HEALTH BENEFITS

Juniper wood and leaves contain diterpenes and other essential oils, which can be antibacterial. The leaves, bark, and twigs also contain six kinds of analgesic flavonoids.

LODGEPOLE PINE
PINUS CONTORTA VAR. *LATIFOLIA*

ROCKY MOUNTAIN LODGEPOLE PINE, TWISTED PINE, BLACK PINE, SCRUB PINE

Family: Pinaceae
Parts Used: whole plant
Season: spring, summer, fall
Region: North America

I was once asked to facilitate a story retreat at Mazama Lodge, near Government Camp, Oregon. It's a beautiful location: the front of the lodge faces the southern flank of Mount Hood, and because it's at 4,200 feet, you can see all the way to the peak. The participants for this retreat were 18 university students in their senior year. They were all preparing to dedicate their graduate studies to climate change,

Lodgepole pines often grow in dense stands.

the ecological sciences, or some kind of conservation research. My job was to help them prepare themselves for what lay ahead by leading them through exercises to help them find out who they really were and what they believed in, and how to let those discoveries guide their future studies and career. We gathered the participants outside the lodge right after they all arrived. They had not really had time to meet each other, so, as a sort of team-building ice breaker, I suggested the group raise a tipi. Afterward, we could all gather inside it for more conversation. I had already brought the

supplies, complete with 20-foot-plus lodgepole pine supports. I gave the group of students a couple of ideas and hints with regard to putting up the tipi and then let them go at it.

Traditionally, it was the Plains Indian women who raised the tipis (lodges) when the clan or tribe arrived at a location where they planned to stay for a while. Historic literature and some firsthand accounts from early explorers and ethnographers suggest that a group of Lakota, Cheyenne, or Omaha women could raise a lodge in 20 to 30 minutes. They could erect an entire village in one full

Plains Indians traditionally used lodgepole pines as the supports for tipis.

afternoon, complete with cooking fires and children running around among the lodges. With Mount Hood looking down on them, my group of college-age students were still working two hours later to get their one lodge raised. Finally, after I gave them a few more hints and suggestions, they got the lodge up, and we all entered the lodgepole pine structure for formal introductions.

USES

The Lakota call lodgepole pine *wazí čháŋ* ("mother pine"). Its straight slender branches and trunks are the preferred source for the

Cheyenne using a horse travois to transport children, circa 1889.

poles inside of tipis. In the past, the poles also provided the frames they strapped onto the shoulders of their horses to fashion a drag sled (travois) for carrying people and provisions from camp to camp. To this day Plains Indians travel on gathering pilgrimages to the Rocky Mountains in order to collect the coveted pines.

Lodgepole pine offers more than a source for tipi frames. The sticky gum can be made into a poultice for treating rheumatism and is applied to the chest for heart problems. An infusion of the gum is taken for tubercular coughs as well as for colds and sore throats. A mixture of the gum and bone marrow can be applied to burns and to treat skin rashes. Its antiseptic and adhesive qualities make it handy for applying to cuts and wounds. The sap or pitch from all species of lodgepole pine comes in handy as a general adhesive and for waterproofing footwear and other clothing.

The young light green needle tips of lodgepole pine can be made into a decoction and used as a wash for sore muscles. It can also be drunk like tea, as a general stimulant and tonic. It is useful as a diuretic; good for stomachaches, colds, sore throats, and coughs; and can be applied to skin infections. An infusion or decoction of the inner bark (or cambium layer) is taken for sore throats, coughs, colds, tuberculosis, stomachaches, and ulcers; as a wash for muscle aches; and as a blood purifier. The inner bark is also eaten; it can be dried and pounded into a flour and made into cakes. In addition, the roots of lodgepole pine contain lots of tannins, useful for tanning hides. The nuts from the cones can be eaten; they have a pleasant and mildly sweet piney taste.

IDENTIFICATION AND HARVEST

Of the four varieties of *Pinus contorta*, the variety that native peoples have preferred for centuries is var. *latifolia*. It is an evergreen tree that grows straight and tall, sometimes up to 150 feet. The foliage and crown tend to be thin and narrow, and that branches do not begin for several feet up the trunk is characteristic. The dark shiny needles grow in bunches of two or three. In April, new spring growth buds emerge; they are egg-shaped, reddish, and resinous. The new growth is usually complete by midsummer. Lodgepole pine cones are about 3 inches long and have prickles on the tips of their scales. They are serotinous: that is, they

Lodgepole pines have cones that are serotinous: they will not open until they have been exposed to the heat of a forest fire.

are closed and require occasional forest fires in order to open and to release their seeds to the forest floor. The cones of the northern California Bolander pine (var. *bolanderi*) are also serotinous. The other varieties do not have these kinds of closed cones.

Lodgepole pines are found throughout the Rocky Mountains and sometimes at higher latitudes and elevations in the northern midwestern regions. In some areas, because it is a pioneer species, lodgepole pine is found in early to mid-succession growth. Lodgepole pines are very common in post-burn areas of the Rockies, where they tend to grow in pure stands. The needles can be collected at any time of the year but are best from midsummer through fall. In most states, a permit is required before collecting younger trees intended to be used for tipi poles. Permits are not difficult to acquire, but be prepared for lots of resins and gums getting on your clothes and hands. The bark is most easily scraped off the trunks using a drawknife.

HEALTH BENEFITS

Lodgepole pine needles contain several piperidine alkaloids, predominantly euphococcinine, which makes them useful as an anticonvulsant (especially), antipyretic, analgesic, antibacterial, antirheumatic, antitumor, and antidepressant. The needles also contain glycosides (zingerone, rheosmin, chavicol, benzoic acid), which add to their antibacterial effects and help to reduce stomach irritation. The bark contains several diterpenes with antimicrobial and anti-inflammatory qualities.

MANZANITA
ARCTOSTAPHYLOS SPP.

Family: Ericaceae
Parts Used: leaves, fruit, bark, wood
Season: year-round
Region: West Coast

One of my first forays into woodworking, at the tender age of ten, involved a twisted branch of manzanita, still covered with its shiny red-brown bark. I didn't want to remove the bark. The branch looked beautiful and spectacular just the way it was. But I knew that eventually the bark would peel away, leaving only

A manzanita in full bloom.

a twisted piece of old wood. I began removing the bark with my fingers and then had to switch to some heavy-grit sandpaper. The hard white inner wood began to reveal itself. I continued to work at the wood using lighter sandpapers until the branch surface was shiny smooth, accentuated with dark elongated knots. I treated the branch with layers of linseed oil, which further brought out the natural grain and beauty of the branch. I presented the piece of wood to my mom, who seemed to really appreciate my handiwork. She honored me, and the wood, by placing it among the family photos and knickknacks on top of the old stereo console we had in our tiny living room. That piece of manzanita was still at my mom's house, last time I checked—50 years of memories later.

USES

Like its close cousin bearberry, manzanita berries are eaten by West Coast tribal peoples. If they are not eaten fresh, they are dried and stored for later use. The Karuk make a tasty cider-like drink from the berries. They can also be pounded into a meal and mixed with other foods, such as wild meats, or made into a mush. The seeds, when separated from the berries and ground, are also made into mush or cakes. Several California Indian tribes apply an infusion of the leaves to the itchy rash that results from exposure to poison oak; the same infusion is applied to sores and wounds. An infusion of the leaves and bark is taken to treat diarrhea. Some California Indians collect the leaves and add them to tobacco smoking

mixtures. The wood of manzanita is extremely dense and strong; it is often made into pipes, walking sticks, cooking tools, hand tools and is used as fuel for bright hot fires and in housing construction.

IDENTIFICATION AND HARVEST

The specific epithets of the various *Arctostaphylos* species reflect how extremely adaptable manzanita is. Generally, manzanita is a woody, erect, multi-branched evergreen shrub. It can grow as a low, 2- to 3-foot-high ground-covering shrub and as a stand-alone shrub up to 8 feet tall. Manzanita's chief characteristic is its cinnamon to dark red bark. The bark of young trees often appears to be peeling, but it becomes smoother as it matures. The dark green foliage is dense with pale gray-green elliptic leaves, 1 to 1.5 inches long. Small urn-shaped white to pink flowers, arranged in dense panicles, open from March to May. Later the flowers produce reddish brown berries that resemble small apples.

Look for manzanita in California's coastal area and the northern coastal ranges, extending eastward and southward to the Cascades and Sierra Nevada foothills. Manzanita does well in clay and sandy soils and thrives with very little water in canyons and on rocky and/or dry drought-stressed ridges and slopes. It will often be found growing in woodlands and chaparral alongside oaks. The leaves are best collected during the spring to summer growth period. The berries can be collected from late summer and on into the fall.

Manzanita is characterized by shiny cinnamon-colored bark and white inner wood.

HEALTH BENEFITS

The main phytochemical constituent of manzanita is arbutin, a glycoside that, when oxidized, converts to hydroquinone, which works as an antibacterial. Manzanita also contains many tannins, which are astringent.

MAPLE
ACER SPP.

Family: Sapindaceae
Parts Used: bark, wood, sap
Season: year-round
Region: Northeast

That maple trees loom large in the origin stories of Algonquian and Iroquoian speakers from the Northeast is not surprising: maple trees are ubiquitous throughout that portion of our continent. The story varies slightly from tribe to tribe, but the general narrative goes like this. During the early days of mankind, there was a cultural hero who acted as a sort of mediator between this world and that of the Creator. Among the Anishinaabe, this figure is Nanabozho, who on occasion would visit the different villages to check up on the people. Life was easy during this time. Game was plentiful, and there was always plenty to eat. Life was so good that sweet syrup flowed freely from the twigs of maple trees. All the people had to do was to break off a twig and pour the sweet happiness directly into their mouths. During one of Nanabozho's visits to a village, he noticed that there was nobody around. Nobody was collecting berries, no one was fishing, there wasn't anyone clearing weeds among their crops. Finally he found everyone in the village in a nearby grove, lazily lying on their backs underneath the maple trees with their mouths open, letting the syrup drip right in.

When Nanabozho saw what was happening, he decided that this could not be allowed

A view up into the canopy of a large old maple. Each maple tree is a reminder to always honor the gifts that the land offers and also of the importance of hard work.

Birch bark basket, designed for gathering maple syrup.

to continue. He constructed a watertight container from some birch bark, filled it with lake water, and returned to where everyone was lounging underneath the trees. He began to pour the water into the tops of the maple trees, diluting the syrup. Very quickly the sap that the people were drinking from the trees became thin and watery; it had little maple

flavor or sweetness. Nanabozho announced that from that day on, if anyone wanted to eat sweet thick maple syrup they would have to gather the thin syrup into birch bark containers like the one he used to carry the water in. They would have to heat stones in fires; they would have to drop those heated stones into the containers of the thin sap in order to thicken it again. In this way, they would learn to be grateful for the gift provided by the maple trees. In addition, they would be able to gather the sap only during the late winter and early spring, so that they would have to pay more attention to their other duties, such as hunting, gathering, and growing their crops.

USES

The preceding story reminds us of the most well-known use for maple—it is the source of a delightful edible syrup. Maple is more than a sweetener, however. An infusion of the inner bark of sugar maple is used for sore eyes and as a cough remedy. The inner bark of silver maple is used for the same purposes as well as being a remedy for hives. The sap of sugar maple is also known to be a blood purifier.

Of course, as for other trees, the wood of maple trees has several utilitarian uses. Maple is considered a hardwood. It is not quite as dense as cherry or walnut, but it is about the same hardness as oak and ash, and harder than other American hardwoods such as hickory, pecan, and poplar. Its dense interlocking grain pattern makes maple wood difficult to work and carve; however, it is this very character that gives maple some spectacular finished surfaces, such as the flame or quilted pattern seen on some stringed instruments.

IDENTIFICATION AND HARVEST

About 12 species and subspecies of maple have been identified throughout North America; this number does not include those introduced from Europe and Asia. Sugar maples (*Acer saccharum*) typically grow 80 to 110 feet tall; they are deciduous and are renowned for their spectacular fall foliage. Silver maples (*A. saccharinum*) normally top out at about 80 feet. Where sugar maples tend to maintain a single tall and stout trunk, silver maples begin

Maple seeds are enclosed in papery "wings" that help them drift on breezes for distribution.

Maples' famous fall colors and iconic palmate leaf shape.

to branch out lower on the trunk. The leaves of both species are palmate and arranged on the stems opposite from each other; the underside of silver maple leaves is silver. Both species have small, inconspicuous bell-shaped flowers that droop down in clusters in conjunction with the emergence of the leaves. All maples produce their seeds in the form of clusters of winged samaras. The bark of young maples is smooth and gray-brown; as trees mature, the bark thickens, forming vertical fissures and ridges that curl out from one edge.

Most American Indian uses for maple are associated with the syrup, which is collected during the late winter and early spring, just when the nourishing sap begins to make its way back up into the tree. Pre-Columbian natives tapped the trees by cutting a V-shaped gash into the trunk using an axe and then collecting the sap in birch bark pails. Later,

Euro-American colonists devised taps and spiles out of ash and other hardwoods and drove them into the trunks of the trees. The thick bark, often easy to separate from the tree, may be collected throughout the year. Most native peoples did not maintain many uses for maple wood due to its density and hardness. It was only with the emergence of modern-day iron and steel tools that maple lumber became favored for tables, cabinets, and flooring.

HEALTH BENEFITS

Pure maple syrup contains several important nutrients, including zinc, potassium, manganese, thiamine, calcium, iron, magnesium, and riboflavin, as well as phenols that are antioxidant, antidiabetic, analgesic, anticancer, and antibacterial.

MESQUITE
PROSOPIS SPP.

Family: Fabaceae
Parts Used: leaves, fruit, roots, bark, wood, gum
Season: year-round
Region: Southwest

The Paiute elder was looking directly at the National Park Service employee, shaking her forefinger back and forth in the air while making her point. We were inside an air-conditioned government building, sitting around several tables, while outside the heat of the Mojave Desert caused distant objects to shimmer in blurry waves. The meeting had

Mesquite inflorescences appear in long cylindrical spikes.

been arranged in order to discuss how to manage local groves of mesquite. The elder continued to scold the ranger about how the Service had allowed the groves to "become a mess." For years, the Service had assumed that leaving the mesquite groves alone would encourage the trees to return to their "natural" ecological state. Therefore, humans—including local native peoples who had relied on the mesquite trees for food, wood, and medicine—had not been permitted access to the mesquite. The result was that the groves had become overgrown with invasive and competitive plants.

Their numbers were dropping—along with that of other native species.

Local Paiute, Mojave, Chemehuevi, and other native peoples have argued for decades that the mesquite groves must be managed by pruning, small-scale controlled burns, and harvesting. In other words, humans play a part in the ecological dynamics of certain plants. When people properly harvest the mesquite, their gathering and pruning encourages more edible pods to grow. Small-scale fire reduces competition from other plants, encouraging healthier mesquite trees and more pods. Open mesquite groves in turn encourage native flora and fauna to remain in the area—natural diversity returns. Everyone wins.

USES

From western Texas to California, desert mesquites have long been a source of food and medicine for several native communities. The hardwood has also been collected for fuel and other utilitarian needs. The fruit pods of mesquite provide an important ancestral food to the Maricopa, Pima, Apache, Hualapai, Cahuilla, Luiseño, Mojave, Panamint, Shoshone, Chemehuevi, Ute, Goshute, Comanche, Navajo, Pueblo, and Southern Paiute peoples. The long pods and their large inner seeds are ground in a stone or hardwood mortar; the result is a sweet, hearty flour that is made into breads, cakes, mush, and thick nourishing drinks. My favorite chocolate chip cookie recipe includes mesquite flour. The seeds are also sometimes parched and ground into a meal for an extra-tasty flour. The pods can also be collected directly from the tree and chewed on for an immediate sweet snack.

The black gum from the mesquite is an important medicine to the Pima. It is boiled with a little water and applied to chapped fingers and sore lips and gums; it is taken internally to cleanse the system. Mesquite leaves are pounded and boiled and placed on the eyes of Pima individuals as a treatment for pinkeye. The Pima use black gum in a concoction to dye gray hair black.

The juice from the leaves of honey mesquite (*Prosopis glandulosa*) is applied to the eyelids to treat irritation. The Comanche chew the leaves of mesquite and swallow the juice as an antacid. The Pima use the boiled root bark of screwbean mesquite (*P. pubescens*) as a poultice for wounds. The root bark is also made into a tea to regulate a woman's menstrual cycles. The gum from mesquite trees is gathered, soaked in water, and used as a cleansing eyewash for pinkeye and to treat sores and skin infections. A poultice of the chewed leaves is used to treat red ant bites and sores. A cold decoction of the leaves is applied to the head to relieve headaches and stomachaches. A decoction of the beans is used to treat sunburn. An infusion of the roots is taken to treat diarrhea.

The Navajo use the strong and flexible wood from honey mesquite to construct bows. Mesquite wood is used across the Southwest as fuel for hot fires and for pit roasting. The wood is used to build homes and for making cooking utensils, toys, traps, fencing, baby cradles, and farming tools. The inner bark and root material of mesquite is used for basketry and for making cordage.

Mesquites are generally medium-sized deciduous trees or tall shrubs. Velvet mesquite (*Prosopis velutina*) can reach up to 45 feet tall; screwbean mesquite can grow up to 30 feet; honey mesquite grows 12 to 18 feet tall. Mesquites tend to create spreading, rounded crowns. All parts of the trees contain short, dense hairs. The branches tend to grow crookedly and sprout half-inch or longer spines that appear at the nodes. The compound leaves are fern-like, divided into many tiny leaflets. The inflorescence of mesquite grows into a spike-like yellowish raceme. The trunk of velvet mesquite has shaggy bark, and its flat beanpod-like fruits are 6 to 10 inches long and tan, sometimes streaked with red. Screwbean mesquite produces clusters of two to ten pods, each 2 to 4 inches long, which are tightly coiled. The thick, reddish brown pods of honey mesquite are the largest of the genus, at 4 to 12 inches long, and contain half-inch-long seeds.

Look for mesquites in the desert regions of the Southwest, from southern California to western Texas. They tend to grow in or near washes, ravines, and arroyos. The leaves, bark, and roots of mesquite can be collected at any time of year. The fruitpods can be harvested in the spring, while they are still green; they can also be collected later in summer, after they have ripened and are the sweetest. Most native peoples collect mesquite pods during the late summer, as this is when the pods are traditionally pounded into flour. As with other spiny plants that grow in the Mojave, Chihuahuan, and Sonoran deserts (e.g., cholla, ocotillo, prickly pear, barrel cacti), be careful of the spines when collecting mesquite pods.

Ripe mesquite pods can be plucked from the tree and eaten directly as a sweet trail snack.

HEALTH BENEFITS

Mesquite contains several phytochemicals, including the alkaloid juliflorine, which is analgesic, antimalarial, and antifungal. Its phenolic compounds are anti-inflammatory and antitumor; its terpenes are antibacterial and anthelmintic. Its quinones have various pharmacological functions, including being antiemetic, anticancer, antidiabetic, and antiulcer.

MILKVETCH
ASTRAGALUS SPP.

Family: Fabaceae
Parts Used: leaves, fruit, roots
Season: spring, summer, fall
Region: North America

Toward the end of every fall and during the winter, Navajo people gather at specific places around the large Navajo Nation that is spread across the Four Corners region. They bring food, gifts, and prayers as part of the Night Chant, a nine-day healing ceremony that is performed to restore order and balance in the Navajo universe and to heal not only some of the people in attendance but also the land, the animals, and the plants. The ceremony also presents an opportunity for young Navajos to become initiated into *hózhó*, the Navajo concept of balance and beauty. Yeibichai (Talking Gods) appear and specific chants are performed during four major sections

of the ritual. During the ceremony, the singers and participants engage in sweat baths, ritual cleansings, masked dance, and the ingestion of medicinal plants. On the last night of the ninth day, the closing chant is performed. The participants recite after the main singer several lines offered to the thunderbird of pollen. The song lines ask the Yeibichai and the thunderbird of pollen that the people be allowed to walk in balance with all nature. At one point, they recite: "May fair plants of all kinds come with you." The song ends thus:

In beauty I walk,
With beauty before me, I walk,
With beauty behind me, I walk,
With beauty below me, I walk,
With beauty above me, I walk,
With beauty all around me, I walk,
It is finished in beauty,
It is finished in beauty,
It is finished in beauty,
It is finished in beauty.

Milkvetch in bloom.

During this most important of ceremonies, the small purple-flowered *Astragalus allochrous* (*txáa'iiltchóciih* in the Navajo language) plays an important role; its leaves are used as a ceremonial emetic. It too is considered a Navajo life medicine.

USES

An infusion of the leaves can be taken as a laxative or as an emetic. The leaves are ground up by the Cheyenne and applied to poison ivy rash, and a poultice of the leaves can be used

Milkvetch flowers can range in color from white to cream to purple.

externally for back pain. The Acoma eat the young root of *Astragalus lentiginosus*, and the Hopi eat the roots of *A. ceramicus*. An infusion of the root is applied topically to sores and wounds and tired, sore eyes; it can also be taken for stomachaches and coughs and to reduce fevers. A decoction of milkvetch root is applied to bleeding wounds and given to people suffering from convulsions; it is also used to reduce fevers, to relieve menstrual pains, and to treat sore throats and toothaches. The roots are often simply chewed, to relieve toothaches, coughs, and sore throats; as a cathartic; to reduce fevers; and to treat stomachaches, the flu, and cramps. Finally, the seeds are collected and either eaten directly or pounded into a flour and used in other dishes.

IDENTIFICATION AND HARVEST

Astragalus is in the bean family, and one of the identifying characteristics of these little annuals and perennials are the greenish white to cream to purple pea-shaped flowers that often grow at the end of long stalks. Following the flowers are smooth, erect, woody fruit pods that contain several tiny seeds. The leaves of milkvetches are smooth and pinnately compound; the leaflets grow up to 2 inches long. Most plants are small, reaching only about a foot in height; a few species (e.g., *A. canadensis*) grow up to 5 feet tall. Around 300 species of milkvetch can be found growing across much of North America, from Vermont to Virginia and west to Washington and California; plants most often occur in open plains and prairie landscapes, and on mesa tops and other open areas of the Southwest.

HEALTH BENEFITS

In general, *Astragalus* species contain galactomannans, saponins, amino acids, flavonoids, isoflavonoids, alkaloids, astragalosides, and terpenes. These phytochemicals demonstrate a variety of pharmacological activities; besides acting as immunoregulators, expectorants, and gastrointestinal protectors, they are antioxidant, diuretic, anti-inflammatory, bactericidal, hypotensive, antidiabetic, hepatoprotective, neuroprotective, and analgesic.

MILKWEED
ASCLEPIAS SPP.

Family: Apocynaceae
Parts Used: whole plant
Season: spring, summer, fall
Region: North America

Reflected in so many American Indian worldviews and cultures are deep understandings and recognition of the cyclical nature of life. My people, the Rarámuri, believe that when we leave this life our *iwí*, our spirit, transforms into a butterfly. We then travel about visiting

Milkweed's flowers, frequently pink to magenta, are clustered in umbels.

our favorite places and people before floating up into the night sky to join the many stars of the Milky Way. It is understandable, then, that species of milkweed are special to my people as well as to many North American Indians, who see butterflies as symbols of the cycles of life. Milkweeds attract several species of butterflies, including the endangered Monarch. The continuing presence of these plants protects our butterflies; they depend upon it for survival. Milkweed nectar is eaten by butterflies, bees, and hummingbirds. To humans it is a food, medicinal, and utilitarian gift.

USES

Approximately 100 *Asclepias* species are native to North America. Here I will focus on showy milkweed (*A. speciosa*) and inmortal (*A. asperula*), but most other species have similar uses. In the Southwest inmortal is dried and snuffed by the Navajo and other native peoples to relieve excessive mucus. It is also used as a ceremonial emetic. The milky latex is boiled until it thickens and then used as a chewing gum. Like other milkweeds, the fibers from inside the large pods are used as a stuffing for pillows and sleeping pads and twisted into cordage. The powdered root is made into a decoction to reduce fevers and pains in the chest and ribs.

Showy milkweed has similar uses and much more. The latex is used to get rid of warts and parasites; the latex also has antiseptic properties and so is used to treat skin sores. A decoction of the plant tops is strained and used to treat snow blindness. The root can be chewed fresh or dried into a powder and used to treat stomachaches. A decoction of the root is taken to treat coughs and bloody diarrhea; a poultice of the roots is applied to the joints to relieve pain from rheumatism and arthritis. The Paiute used a decoction of the seeds to draw venom from snakebites. The latex is boiled and mixed with animal fat for a chewing gum. The inner parts of the young stems as well as the young pods are boiled and eaten or simply eaten raw; they have a flavor that is similar to pea pods. The flowers, which are also edible, are added to soups and stews. The stems do contain fibers, which can be pounded into a temporary cordage. The Pomo use the fibers from the pods to make clothing and cordage.

IDENTIFICATION AND HARVEST

Inmortal often grows as a low, creeping plant, whereas showy milkweed is an erect herbaceous perennial, up to about 3 feet in height. Milkweed's large velvety opposite leaves are borne on short petioles on a smooth single stem. The 5- to 6-inch-long leaves are oblong and cordate at the base. During the spring, milkweed produces clusters of flowers that resemble hooded pouches. The flowers often attract bees and butterflies; they are usually showy, dull purple and pinkish to light green, and borne in clusters either at the ends of the stems or in the upper leaf axils. Each flower eventually produces one or two large warty fruit pods with a seam along one side; this pops open when the pod becomes ripe and dry to reveal several hundred flat brown seeds arranged like scales on a fish. Each seed is surrounded by fine silky white fibers that act like parachutes when the seeds are carried away on the autumn winds. The thick yellow-brown root is hard and knotty, covered with tough bark. It grows deep into the soil and is covered with several lighter brown rootlets.

Showy milkweed and other milkweeds are seen in open fields, disturbed places, and along roadsides throughout North America;

Asclepias asperula is identifiable by its inconspicuous light green flowers.

Within each milkweed fruit pod are flat brown seeds, overlapping each other like fish scales; the fiber from the interior of the pod is used to make clothing and cordage.

HEALTH BENEFITS

Asclepias species contain toxic resinoids and glycosides, which can have a stimulating and lasting effect on the heart, as well as terpenoids and steroidal compounds. Together these phytochemicals provide subtonic, diaphoretic, alterative, expectorant, diuretic, laxative, carminative, antispasmodic, antipleuritic, stomachic, astringent, antirheumatic, antimicrobial, anticancer, cardiovascular, analgesic, and antipyretic pharmacological effects.

MORMON TEA
EPHEDRA SPP.

Family: Ephedraceae
Parts Used: stems, twigs, fruit
Season: year-round
Region: Southwest, Mountain West

Before this plant became known as Mormon tea (or Brigham tea, or Mexican tea)—even before it was classified in the genus *Ephedra*—the Paiute called it *tsurupe*, the Shoshone *durumbe*, the Navajo *tl'oh'azihii*, and the Hopi *hishab*. In fact, it is such an important and useful plant that the Hopi created a katsina doll in its honor. To most people outside the Hopi cultural world, katsinas are small hand-carved wooden dolls purchased by tourists at exorbitant prices. They are supposed to represent Hopi spirits, and indeed, the Hopi do recognize numerous spirit beings, each one of which plays a role in the functioning of the natural world. Throughout the year these beings

they prefer to grow in sandy, loamy, moist soils. Inmortal is endemic to the Southwest; it is often found in sandy, calciferous, rocky soils. Milkweeds can be harvested from spring through fall; plants disappear during the winter months. The milky sap of the leaves and stems is not toxic, but it will stick to the skin and clothing. Please collect sparingly, as the milkweed are important hosts for pollinators.

The stems and twigs of this jointed plant are steeped into a medicinal tea.

USES

Many native peoples of the Southwest steep the greenish twigs and stems of green Mormon tea (*Ephedra viridis*) for a beverage that is stimulating and aromatic, reminiscent of mild earthy pine. Mormon tea bears seeds at the joints of the plant stems; some species (e.g., *E. viridis, E. nevadensis*) bear enough of the

assume semi-human form during rituals and ceremonies. As a way to teach about the various spirits, wooden representations (*katsina tihu*) are carved and given to children. With the advent of the first early ethnographers and then tourists, these carvings became collector's items. There are katsina for all parts of the natural world, including rain, animals, clouds, insects, and, of course, plants. In addition to their hishab katsina for Mormon tea, the Hopi have plant (*tusak*) katsinas for prickly pear (*navuk'china*), squash (*patung*), and beans (*muzribi*), among others. Plant katsinas are very important to the Hopi: it is felt that, because plants require water in order to survive, plant katsinas must be directly connected to that life-giving substance. This is extremely relevant on the Hopi homeland of the Colorado Plateau, where rain is scarce.

A Hopi katsina depicting the personification of the Mormon tea plant. This hishab katsina was created by Josie Seckletstewa.

seeds to be collected and ground into flour for a mush-like dish. The seeds can also be roasted and then ground into a coffee-substitute drink. The stems and twigs are also made into an infusion by the Shoshone and Hopi and taken to treat gonorrhea and syphilis. The same infusion is used to clear bladder and kidney infections. The Washoe give an infusion to women in order to relieve difficult menstruation. Other California Indians take an infusion of Mormon tea for backaches. The Tewa make a decoction of the ground twigs and use it to relieve diarrhea. A poultice of the twigs is applied to sores and burns, and an infusion of the twigs is used to treat coughs and colds, rheumatism, and stomach ulcers. I have been in sweat bath ceremonies when Mormon tea twigs are steamed on the central coals in order to purify and cleanse the participants. It is believed that consuming Mormon tea at the change of seasons helps to prevent illness.

When Mormon tea is steeped in hot water, the resulting tea is darkly greenish tan. It is not surprising, then, that the Navajo use Mormon tea for a light tan dye. The Kawaiisu insist that charcoal from *Ephedra californica* is the best for tattooing.

IDENTIFICATION AND HARVEST

Twelve *Ephedra* species are found in North America. In general they are 3- to 5-foot evergreen shrubs with yellowish gray, erect-spreading branches. Often *E. viridis* will appear more green in contrast to others in the genus. Mormon tea is an ancient plant that has been around for about 200 million years. It is more closely related to conifers than to herbaceous plants. The plant does grow very small paired leaves, but they fall off

Yellow-brown flowers form into cones at the joints, revealing that Mormon tea is more closely related to conifers than to herbaceous species.

the plant very early during its growth stage, which begins in the early spring. The result is an apparently leafless plant with a jointed, broom-like appearance. Tiny, yellow-brown clusters of flowers form into cones at the joints; as they mature, these cones produce tiny seeds. Mormon tea occurs in high desert regions at 3,000 to 7,000 feet elevation, from Texas and Oklahoma northwest to Oregon. It will be one of the few green-looking shrubs, growing in contrast to the browns and tans of the arid basins, sandy slopes, and rocky hillsides of western North America. It is found in the California oak woodlands, in the Joshua Tree area, in juniper-piñon woodlands, and in the Sonoran and Chihuahuan deserts, in company with such plants as sagebrush, saltbush, and creosote bush. Mormon tea can be collected any time of year, but it is best when in its flowering stage.

HEALTH BENEFITS

The scientific consensus is that only *Ephedra sinica* and other Chinese species contain enough of the alkaloids ephedrine and norephedrine to be pharmacologically effective. The North America species do not contain alkaloids; however, they do contain other pharmacologically active constituents, including flavonoids, tannins, and polysaccharides. These phytochemicals are anti-inflammatory, anticancer, antibacterial, antioxidant, antiviral, diuretic, and hepatoprotective; they also show antiobesity activity. The tannins, which are astringent, help treat coughs and colds by reducing the secretion of mucus.

MOUNTAIN LAUREL
KALMIA LATIFOLIA

SHEEP KILL LAUREL, CALICO BUSH, IVY BUSH, MOUNTAIN IVY, SPOONWOOD

Family: Ericaceae
Parts Used: leaves, flowers
Season: spring, summer
Region: Northeast, Southeast

Be careful not to confuse mountain laurel with *Sophora secundiflora* (Texas mountain laurel, mescal bean). Both plants are poisonous and both are sacred to American Indians, but they are used very differently and inhabit different ecosystems. In the eastern half of the United States, mountain laurel is better known as spoonwood, a common name bestowed by early colonists, who noted the beautiful everyday implements that some indigenous peoples had carved from its wood.

From a distance, mountain laurel resembles a rhododendron.

USES

Although deer have occasionally been observed foraging on mountain laurel, the plant is toxic to humans, and medicinal applications are therefore primarily external. The Cherokee make an infusion of the leaves and apply it to relieve the pain of some deep scratches. The leaf infusion is also rubbed on the skin to relieve rheumatism, as a disinfectant liniment, and to rid the body of parasites and inflammatory fevers. The leaves and flowers of mountain laurel release a secretion that is used externally to prevent muscle cramps. The Cree take a decoction of the leaves as an antidiarrheal.

IDENTIFICATION AND HARVEST

Mountain laurel is a bushy, rhododendron-like woody shrub, 7 to 15 feet tall. In low light, it can resemble a small tree with a single trunk. Its simple alternate leaves are thick, leathery, and glossy. The leaves are 2 to 5 inches long, elliptical, and evergreen, dark green above, yellow-green below. Young growth is often yellow-green. During the spring, mountain laurel produces large clusters of pinkish white flowers that nearly cover the entire shrub. The flowers grow in terminal clusters. Each flower has five lobes centered by ten anthers. The flowers darken over the summer and give way to small fruits that persist into the winter. The bark is thin, smooth, and dark red-brown; it shreds and splits as the tree ages. Mountain laurel is native from New England south to the Florida panhandle, west to Louisiana, and north to southern Indiana. It grows in the rocky and open parts of cool mixed hardwood forests and meadows, the bald tops of mountains, and on mountain slopes. The shrubs often do well on northern slopes and steep hillsides near wet areas. Remember that all parts of the plant are potentially toxic; keep away from pets and children.

HEALTH BENEFITS

Twelve different flavonoids have been isolated in mountain laurel leaves. Flavonoids are known to be antiviral, anti-inflammatory, and antibacterial. The leaves also contain five kinds of grayanotoxins, which, despite their toxicity, are analgesic, astringent, disinfectant, narcotic, sedative, and antifungal.

MOUNTAIN MAHOGANY
CERCOCARPUS SPP.

Family: Rosaceae
Parts Used: leaves, bark, wood, roots
Season: spring, summer, fall
Region: western North America

Mountain mahogany is one of my favorite plants. Whenever I think of these scraggly-looking shrubs, I am reminded of time spent traversing the middle elevations (4,000 to 7,000 feet) of the Mountain West and Southwest uplands, where the greatest plant and animal diversity resides. These upland habitats of the West are where many American Indians spent much of their time growing food crops, hunting, and wildcrafting edible and

The plume on *Cercocarpus montanus* is its pistil.

medicinal plants. Mountain mahogany thrives at these temperate zones and offers many uses for American Indians.

USES

Three species of mountain mahogany are widely used: alderleaf mountain mahogany (*Cercocarpus montanus*), curl-leaf mountain mahogany (*C. ledifolius*), and birchleaf mountain mahogany (*C. betuloides*). They afford similar uses as medicines, as materials for ceremonial and utilitarian items, and as dyes. The leaves and bark of alderleaf mountain mahogany is made into an infusion by the Tewa and taken as a laxative. The Navajo use the same infusion for stomach problems. The Kawaiisu take a decoction of the roots as a cough medicine. The western Apache burn the wood and apply the resulting charcoal to the skin as a burn dressing. The root bark of alderleaf mountain mahogany is part of a mixture of other barks used to make red dye for buckskin and cloth. The hard wood makes good arrow points, weaving tools, and handles, and the roots are used to make pipe bowls. The Hopi use the wood to construct prayer sticks, which they leave at specific shrines.

The Shoshone treat tuberculosis with a decoction of curl-leaf mountain mahogany bark. The charred wood of curl-leaf mountain mahogany is also used as a burn dressing and for cuts and wounds. Curl-leaf mountain mahogany exudes a kind of resin that is dried and used for earaches. An infusion of the bark is used to treat stomachaches and diarrhea and for colds and coughs. A decoction of the leaves and bark is used to treat stomach ulcers, heart problems, diarrhea in children, pneumonia, colds, and eye diseases. The wood of curl-leaf mountain mahogany is made into durable sticks for digging edible tubers and for the tips of fishing sticks. The Pueblo of Jemez mix the bark with alder bark to make a red moccasin dye. Birchleaf mountain mahogany is similarly employed for all uses.

Cercocarpus ledifolius in the rugged habitat of the Steens Mountain Wilderness, Oregon.

IDENTIFICATION AND HARVEST

Mountain mahoganies are medium to large shrubs or small trees, erect, with many spreading branches. Alderleaf mountain mahogany grows to 20 feet tall in ideal conditions; curl-leaf and birchleaf mountain mahoganies can reach heights of 30 feet and often have multiple trunks. The leaves of mountain mahogany are most often dull green above, whitish beneath. The small clusters of simple alternate leaves grow on short spur-like branches; they are deeply veined underneath. Leaf shape varies among the species. The leaves of alderleaf mountain mahogany have an entire margin until the final third, where they become coarsely dentate with two to four ovate teeth. Curl-leaf mountain mahogany leaves are finer, smooth-edged, oblong, leathery, and tipped at the end. The leaves of birchleaf mountain mahogany are also smooth until about halfway, where they become toothed, ending in a rounded birch-like tip. Mountain mahogany flowers from April to June in Arizona and hotter regions, from May to November elsewhere in the Mountain West; a plume-like curvy pistil, 1 to 2 inches long, protrudes from each tan to yellowish tubular flower. The root systems of mountain mahogany are extensive.

Cercocarpus species are found from western Texas to California and north to Montana. Their preferred habitats are sunny dry areas of foothills, mixed oak forests, chaparral, and piñon-juniper communities at 4,000 to 7,000 feet. The best time to gather mountain mahogany bark, leaves, and wood is from late spring to early fall. A scraping tool such as a knife blade might be useful for collecting the bark.

HEALTH BENEFITS

The leaves and bark of mountain mahogany contain dhurrin, a potentially toxic cyanogenic glycoside that in small doses serves as a muscle relaxant useful for treating and soothing dry coughs. Mountain mahogany also contains tannins, which act as astringents.

NAHAVITA
DICHELOSTEMMA CAPITATUM

BLUE DICKS, BRODIAEA, WILD HYACINTH, CONGESTED SNAKE LILY

Family: Asparagaceae
Parts Used: flowers, roots
Season: spring, summer
Region: West Coast, Southwest

Every native culture in North America was reliant on what their local habitats provided in the way of natural resources. Each landscape has its own unusual or endemic food staples. On the eastern lee of California and Oregon's coastal ranges, Paiute, Shoshone, Cahuilla, Karuk, Miwok, Pomo, and other native peoples were given the nickname "diggers" by white intruders. It was a derogatory moniker but one that reflected how many native peoples gathered an important food staple: nahavita tubers. By the middle of spring, the nahavitas had emerged and bloomed, tinging the many

The simple, purple-blue flowers of nahavita are beacons, showing native peoples where to dig to find the tubers below.

rangelands and hillsides in California and Oregon with purple. The flowering of nahavita plants was a cue. Armed with digging sticks, the native western peoples scoured the hillsides and ranges for the tasty little tubers growing underneath the plants. But with the introduction of nonnative and invasive species, and changes in how the landscape was managed by first the Spaniards, then the Mexicans, and now the United States, nahavita has become difficult to find. Nevertheless, whenever I am out on a hike in the hills near my Bay Area home, I keep my eyes peeled for the bright little purple flowers, peeking their heads up among the grasses next to the trail. Occasionally, I spot a nahavita and offer a thought of gratitude.

Although nahavita is primarily harvested for its edible tubers, the Apache and other tribes eat the young flowers as a snack. The corms (tubers) can be eaten raw but are more often roasted, baked, or boiled and then eaten. The corms are also dried and ground into a flour to be added to stews and soups, or made into cakes and breads. In the past the corms were stored for use after the harvest. The Kawaiisu have an interesting use of nahavita corms: they rub them on a stone, forming a sort of glue that is then spread onto seed-gathering baskets, effectively sealing the small openings between the weaves.

Nahavita is an herbaceous perennial. The main visible characteristic is a fleshy central stem ending with a dense cluster of purple-blue flowers. Each flower has six stamens. The central stem will sometimes twist around nearby plants. The plant has a set of several narrow basal leaves. Underneath the plant are several yellowish white cormlets covered with a light brown husk. Nahavita does well in open grassy habitats, open hillsides, mixed evergreen and deciduous forests, oak woodlands, and chaparral. It often emerges around vernal pools and in open disturbed areas and post-fire woodland environments. It is native to California, Oregon, and parts of Arizona and New Mexico. Harvest the tubers beginning in the middle of spring and into the summer. A digging trowel is recommended. The tubers should be cleaned of dirt and rinsed with water before eating.

Nahavita is a good source of complex carbohydrates, soluble fiber, nutrients, minerals, and protein.

OAK
QUERCUS SPP.

Family: Fagaceae
Parts Used: leaves, nuts, bark, wood
Season: year-round
Region: North America

For American Indians across North America, oaks provide a fountain of uses, from materials for tools, construction, toys, dyes, even games, to medicine and food. I will never forget the first time I tried venison and acorn stew. It is a delicacy on the White Mountain Apache reservation in eastern Arizona, but it

The leaves and acorns of *Quercus garryana*.

Acorn granaries at a Miwok village in California, circa 1877. These were constructed above the ground and were made of woven grapevines and buckrush. They were lined with wormwood and shingled with conifer boughs.

was a new taste for me: my people transform acorns into mush and meals, not stews. On the White Mountain Apache reservation the dish is served during ceremonies and at other special, celebratory occasions. I was at a local festival, where a vendor was selling styrofoam bowls of the venison-acorn stew. My first reaction—a greasy, lard-like sensation in my mouth. It was not unpleasant, just unexpected. This was followed by the gamey taste of venison coupled with the slightly bitter, earthy flavor of the acorn. I loved it.

USES

Oak trees are native throughout the continent and were utilized by nearly all American Indian cultures, so this section will necessarily be somewhat generalized, offering specific tribal examples where appropriate. Acorns, the nuts of oak trees, are eaten by American Indians across the continent. The acorns are normally ground into a fine meal and either used to make breads, mush, and porridge or added to stews as a thickener; in the case of eastern red oaks, they are also pressed for their oils. Although all acorns are edible, most should first be leached in water in order to remove the bitter tannins. Some acorns contain fewer bitter tannins than others: in the east, white oak (*Quercus alba*) is said to produce the sweetest acorns; people from the South favor pin oak (*Q. palustris*), and in the Southwest it is Emory oak (*Q. emoryi*). Bur oak (*Q. macrocarpa*) in the Midwest is supposed to be very sweet. Native peoples in California treasure valley oak (*Q. lobata*) for its large sweet acorns; it is a kind of white oak but produces acorns so low in bitter tannins that they are sweet enough to eat without any leaching or other preparations. In California, acorns were treated as a currency; they were often traded for other foods or services, like those of a medicine person.

Oaks provide many other uses. The acorns can be used as bait in fish traps. The Cahuilla of California string the dried acorns as a form of jewelry; dried acorns are also gathered on a cord strummed across the teeth as a musical instrument. Oak wood is strong enough for making hunting and agricultural implements, bows, furniture, toys, battens for looms, and many other items where hard wood is needed. Oak is also an excellent fuel wood for cooking. I have made clothing dye from the bark and leaves and use the bark as a tanning agent for deer hides.

Oak trees are a source for several medicines. Both the leaves and bark can be used to reduce bleeding and to treat hemorrhoids; they are also rubbed directly onto the skin to treat sores and insect bites. An infusion of the bark is taken to reduce fevers and to treat dysentery, diarrhea, arthritis, indigestion, mouth sores, sore throat, asthma, and colds; it is also used as a cough medicine, disinfectant, and emetic.

IDENTIFICATION AND HARVEST

The only commonality among oaks is that they are deciduous and bear acorns. Oaks vary in size; they can grow to be large trees, up to 100 feet tall, with broad rounded crowns. One such is white oak; as with other large oaks, it has bright green six-lobed leaves that grow up to 6 inches long. Northern red oak (*Quercus rubra*) is another large oak that grows up to 90 feet tall; the lobes of its large leaves terminate in sharp points. Gambel oak (*Q. gambelii*) of the Rocky Mountains and other parts of the interior west can grow up to 50 feet but is typically no taller than 30 feet; it is more of a tall shrub that sometimes forms thickets, and its deeply lobed, leathery leaves are bright green above, pale below. Toward the other end of the spectrum is canyon live oak (*Q. chrysolepis*); these small trees or large shrubs normally grow 18 to 30 feet tall. The flat, toothed leaves of canyon live oaks are oblong, not lobed, and only 1.5 inches long.

Fifty-eight *Quercus* species occur in North America. Oak-hickory hardwood forests once dominated the eastern portion of our continent, and in California, oak trees played an essential role in the ecosystems and habitats of the Sierra foothills all the way to the coast. Oaks

The elongated acorns of the *Quercus lobata*.

typically flower in the spring and produce their acorns from August to October. The bark can be collected at any time. Native peoples usually collect acorns from the ground, around the base of the selected trees; occasionally people can be seen beating the lower branches of trees to release their acorns. If the shell has a tiny hole, the acorn must be discarded; the hole indicates that an oak weevil has bored into the acorn. The acorns are shelled to reveal the meaty insides and leached to remove the bitter tannins. The shelled acorns are placed in a large pot of water and brought to a boil. The water will turn dark. The process should be repeated until the acorn meat is no longer bitter, at least five times. Another method is to

A Pomo woman cooking acorns in California, circa 1924.

grind the meat and place it in a large jar. The jar is filled with water and then shaken once or twice a day until the bitterness is gone.

HEALTH BENEFITS

Oak bark and leaves are high in tannins and quercetin. Both constituents are pharmacologically active flavonoids and highly astringent, making them useful as antidiarrheals, hemostatics, and antihemorrhoidals. Tannins have also been reported to have antibacterial activity. Quercetin demonstrates anti-inflammatory activity.

OREGON GRAPE
MAHONIA SPP.

Family: Berberidaceae
Parts Used: leaves, stems, fruit, roots
Season: spring, summer, fall
Region: western North America

For me Oregon grape is an associative plant: I associate the low-growing *Mahonia repens* (aka holly grape, creeping barberry, creeping holly) with the arid piñon-juniper woodlands of New Mexico, northern Arizona, and the lower elevations of western Colorado. Despite the generic common name (most often used in reference to the larger *M. aquifolium*), Oregon grapes are not related to *Vitis vinifera*, the common wine grape. The two species that offer the most in the way of medicinal uses are *M. repens* and *M. fremontii*.

USES

The little ripe purple-blue berries of Oregon grape are edible, but most of the time the berries are rather tart. The berries of *Mahonia fremontii* are less bitter than those of *M. repens*, but neither berry is truly sweet or flavorful. Blackfoot, Cheyenne, Hualapai, and other native peoples throughout the Mountain West eat the berries raw and sometimes roasted. The fruits can be crushed and made into a drink as well. The roots of Oregon grape provide the most medicinal value. An infusion of the roots is used to help encourage the delivery of the afterbirth after labor; the same infusion is applied topically to boils and as an antiseptic for cuts and wounds. An infusion or poultice of the roots is used to relieve rheumatism. The infusion can be taken to slow down bleeding

Mahonia aquifolium in flower.

and for dysentery, coughs, and kidney troubles. A decoction of the roots is taken to stop internal hemorrhaging and for stomachaches, kidney problems, and headaches; it is also used as a wash for body aches. The berries can be crushed with water and taken for coughs and as a kidney cleanser. The Navajo use a decoction of the leaves and twigs for rheumatism and as a ceremonial emetic. The roots and stems are also crushed and made into a bright yellow dye; the yellow is a result of the bitter alkaloid berberine.

IDENTIFICATION AND HARVEST

Mahonia repens is a creeping woody evergreen shrub, about 15 inches tall; *M. fremontii* is much larger, up to about 6 feet in height. The leaves can appear to have a waxy coating; their upper surface is often glossy, the lower surface, dull. They are pinnate and ovate to lanceolate, arranged alternately on the branches; they generally grow to about 3 inches long and have spiny holly-like teeth around the edges. Plants produce several clusters of yellow flowers in spring. During the summer, small round to ellipsoid waxy green berries emerge; toward the end of the summer, the berries, although they keep the waxy coating, will darken to blue and purple. The evergreen leaves too darken over the winter.

Look for Oregon grape in the lower and drier foothills of the Mountain West from Montana south to New Mexico and west to the Pacific. The plants often grow in partial shade in rocky areas in and around piñon-juniper woodlands. The best time for harvesting the root and stems is during the fall, after the plant has had the entire growing period to gather and store its beneficial compounds.

HEALTH BENEFITS

Berberine, a benzylisoquinoline alkaloid, is the main phytochemical constituent of Oregon grape; it is mostly concentrated in the roots and stems. Oregon grape leaves contain the alkaloids isocorydine, corydine, thaliporphine, and glaucine. Alkaloids are known analgesics, antibacterials, and cough suppressants; berberine also demonstrates antitumor activity.

The leaves of *Mahonia repens* 'Rotundifolia' are slightly more oval and look less like holly leaves than those of *M. fremontii*.

OSAGE ORANGE

MACLURA POMIFERA

BOIS D'ARC, BODARK, HEDGE APPLE, HORSE-APPLE, NARANJO CHINO

Family: Moraceae
Parts Used: fruit, roots, bark, wood
Season: spring, summer, fall
Region: North America

There was a time when the central to southern Great Plains of North America were populated by mammoths, mastodons, ground sloths, and other large herbivores. These animals ate, among many things, the large strange-looking fruits of Osage orange. They did not pull the fruits from the trees but rather waited until they fell onto the ground. When these animals disappeared, the Great Plains fell under the management of large herds of buffalo, elk,

Each Osage orange is actually a dense cluster of hundreds of small fruits.

antelope, and humans. Unfortunately, these creatures do not eat Osage orange fruits, and with their seeds no longer dispersed by roaming herds of fruit-eating animals, Osage orange's range was temporarily reduced to a very small area of the southern Plains marked by the drainage area of the Red River in east-central Texas, southeastern Oklahoma, and nearby Arkansas. The tree got its name from the tribe of native peoples who primarily occupied this region: the Osage. Fortunately for the tree, its unusually colored wood is very durable and resistant to rot. The wood was so valued by American Indians that bow makers and wood carvers traveled hundreds of miles just to collect it. Osage orange wood was also an important trade item throughout the continent, and it even has some medicinal value. Colonists soon learned the value of Osage orange and, as a result, cultivated it as they migrated. Osage orange is now naturalized in 30 states and introduced in the province of Ontario.

USES

Osage orange is revered primarily for its wood, which is the favored material for making bows. Even Tewa bow makers from the Rio Grande region of New Mexico travel hundreds of miles east into the plains to gather the wood. It is also a favored wood for ceremonial rattles and rasps, peyote staffs, and other items where strong but flexible wood is required. Pueblo and Plains Indian weavers appreciate the inner wood and bark for the yellowish color it imparts as a dye; it is used to color fabric and buckskin. The Comanche and other Plains

Indians make a decoction of the root to be used as a wash for sore eyes. The fruits are used as a natural insecticide and fungicide.

IDENTIFICATION AND HARVEST

Osage orange is often a large deciduous shrub but in ideal conditions can make a medium-sized tree, 30 to 50 feet tall. The branches can span as far as 40 feet from the central trunk and sometimes bend down nearly to the ground. The trunk is usually short and often divides into several large limbs; the grayish bark is heavily furrowed. The alternate leaves are a glossy dark green, smooth on top, pale underneath; they are up to 5 inches long and about 2 inches wide, ovate to lanceolate and rounded at the base. Fall color is yellow-orange. Globular white flowers with large yellow anthers appear in May and June, after the leaves. The distinctive ball-shaped Osage oranges, which appear in September, are slightly bigger than oranges (about 5 inches in diameter) and yellow-green at first. The outer skin of the fruits is rough and deeply wrinkled-looking; inside, several oval seeds

The dark and furrowed bark of a mature Osage orange.

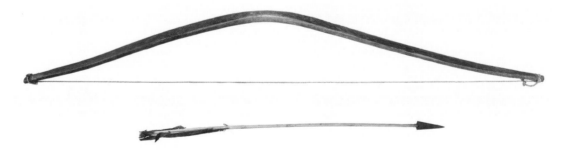

A Sioux bow of Osage orange from Minnesota, first half of the 19th century.

are embedded in the flesh. Look for Osage orange in open woods, fields and thickets of bottomlands, and in and around old farmsteads. When collecting the fruits, be careful of the sticky latex; it can cause dermatitis in sensitive individuals. The wood is very hard and can dull a saw or axe.

HEALTH BENEFITS

The wood contains isoflavones, including osajin and pomiferin, which are strong antioxidants and antimicrobials, and tetrahydroxystilbene glucoside, which demonstrates antifungal and antibacterial activity. Isoflavones extracted from the fruit are also antimicrobial.

OSHA

LIGUSTICUM PORTERI

———

BEAR ROOT, CHUCHUPATE, PORTER'S LOVAGE, COLORADO COUGHROOT, INDIAN PARSLEY

Family: Apiaceae
Parts Used: leaves, roots
Season: summer, fall
Region: western North America

My people, the Rarámuri, believe that everything around us is alive. Or, as native writer N. Scott Momaday puts it, everything around us has "being-ness." We believe that we share breath with all nature, including all the plants, and that we are directly related to everything in nature. We even go so far as to categorize

Osha is one of the remedies most widely used by American Indians.

our plant relatives by ethnicity and gender. Some plants have special status: to us, osha is Wásia ("grandmother").

Osha root is one of the most widely known and used plant remedies by American Indians. During pre-Columbian times, it was a popular trade item among indigenous peoples in North America. Even now, osha root is traded at pow wows and other events where native peoples gather. After contact between Spanish colonists in the Southwest and northwestern Mexico, osha root became a staple remedy among Hispano herbalists. One of the most widely used common names for this plant among American Indians is bear root. Navajo people believe that the plant is a gift from Bear. The root is also held in high esteem among the Hopi, whose Bear Clan is in charge of its collection. There are numerous anecdotal reports of bears in the Mountain West seen chewing the leaves of osha, digging up and chewing on osha root, and looking for osha root immediately after hibernation. Many American Indians will include a piece of the root in their medicine bundles, where it is believed to give protection.

USES

Among the Rarámuri, osha root is highly valued for its powerful healing qualities. Among our healers, the *owerúame*, osha is a favorite. An infusion of the root and sometimes the leaves is consumed to treat gastrointestinal

Osha root is known for its pungent aroma, which is reminiscent of slightly burnt celery root.

ailments, colds, influenza, coughs, and sore throats. The root is sometimes chewed for the same purposes. A wash is made of the root and applied to the body to help alleviate headaches and fevers. The root can be crushed, boiled, or used as a wash to treat rheumatism and as a tea for coughs and pneumonia. The root is also used as a ritual cure during the *yúmari* dance, which is performed to promote the health of the land and everything that lives on it. Osha root is often worn around the neck to ward off sorcerers and snakes.

IDENTIFICATION AND HARVEST

Care must be taken when identifying osha. The aboveground portion of the plant resembles many other plants in the Apiaceae, such as angelica, cow parsnip, and Queen Anne's lace. But most concerning is that this species closely resembles poison hemlock (*Conium maculatum*) and water hemlocks (*Cicuta maculata, C. bulbifera, C. douglasii, C. virosa*), all of which are highly poisonous. With osha the long, deeply dissected basal leaves, three to five in

Osha leaves resemble those of other plants in the same family, some of them highly poisonous; be very careful when attempting to identify osha in the wild.

number, are pinnate and roughly deltoid; the pinkish white flowers are borne in flat-topped umbels. A distinguishing characteristic of osha is the scent of its root: it is very aromatic and pungent but not unpleasant, something like the fragrance of slightly burnt celery root. This is in direct contrast to the odor given off by the roots of its poisonous cousins, which has been compared to the smell of a dead mouse. Osha root is also very dark, fibrous, and "hairy," whereas the roots of the dangerous plants are lighter in color.

Another way to distinguish osha from its poisonous cousins in the Apiaceae is with its habitat. Osha is a high-elevation plant that prefers wet mountainous areas from the northern Sierra Madre mountains of Chihuahua, Mexico, through Arizona, Utah, Colorado, New Mexico, up to Wyoming and Montana; *Conium* and *Cicuta* species tend to grow at lower elevations in these regions. In Colorado, collectors do not begin to search for osha until they have reached the open slopes, small creeks, and moist meadows at 7,000 feet; and more often,

Osha favors high elevations and moist soils.

the plants will not be found until about 9,000 feet, among the aspens and conifers. Osha will grow in small stands among other species of high-elevation flowering and herbaceous plants. It is the root that is sought after. Use a small trowel to dig for the roots, which will be situated only a few inches below the surface. Refrain from harvesting the entire root. Leave some for future growth.

HEALTH BENEFITS

Osha root contains, among other compounds, water-soluble coumarin, coniferyl ferulate, and ligustilide, an essential oil. Coumarin is analgesic, antispasmodic, and anthelmintic. Coniferyl ferulate demonstrates powerful anticancer, antibacterial, vasodilating, and antioxidant activity. Ligustilide is sedative, anti-inflammatory, and antifungal.

PEYOTE

LOPHOPHORA WILLIAMSII

———

HIKULI

Family: Cactaceae
Parts Used: whole plant
Season: spring, summer, fall
Region: Southwest

We had just completed an all-night Native American Church meeting on the Southern Ute Indian Reservation. I ducked through the opening of the tipi and was greeted by a new dawn on the eastern horizon—and one of the

Peyote is exclusively a plant of the southern Chihuahuan Desert.

ceremonial assistants, who offered me a glass of cool water and some dried venison mixed with currants and piñon nuts. It was a tasty way of breaking the fast after a ritual night of chanting, drumming, storytelling, and taking the "medicine," which is how peyote is often described. I moved away from the tipi and joined a Navajo gentleman who was pulling a large mason jar out of the cab of a pickup truck. I recognized him from the ceremony; he had been sitting over on the other side of the lodge. He began to drink a greenish liquid from the jar. I asked him if he liked green tea. He chuckled and told me that he was drinking peyote tea. I reminded him that we had just been eating peyote for the past nine hours. He said he drank peyote tea every day in order to keep his diabetes in check. Since he began drinking the tea, he told me, he no longer required insulin injections.

Cultural renewal sometimes leads to new and innovative approaches to survival. One such example, begun on the southern Plains

Quanah Parker, founder of the Native American Church.

the ingestion of peyote, a spineless cactus. My people and others, such as the Huichol, see peyote as a relative and also a spiritual being. Remnants of peyote found in caves in southern Texas and Arizona have been carbon-dated to 6,000 years ago. Interestingly, these old samples, some from as far north as the Grand Canyon, are slightly different; they do not resemble our modern peyote, which is now found only in the lower, hotter regions of the Chihuahuan Desert. The ancient peyote was larger and more domed in comparison to modern populations of peyote. Navajo oral traditions and those of other Southwest tribes indicate that the region once hosted a peyote variety that grew at high altitudes and was cold tolerant. Recent botanical research suggests that our species may actually be a hybrid of the high-altitude *Lophophora brackiii* and *L. diffusa*, both of which still occur in Mexico.

The Native American Church (NAC) is a relatively new religious denomination that practices peyotism. It originated in Oklahoma and is now the most widespread federally recognized indigenous religion among American Indians. The Comanche leader Quanah Parker, the founder of the NAC, had suffered a gunshot wound during a battle with U.S. federal troops. As part of the healing process, he was given peyote by a Mexican Indian healer, and during his recovery, he reportedly experienced a vision of Jesus. He adopted peyotism (or what he later called "the peyote road") and created the NAC sometime around 1890 (the church was formally incorporated in 1918), borrowing some peyote ceremonial rituals from Mexican Indians and mixing them with contemporary

during the 1890s, remains a testament to American Indian spiritual resilience—but it has also spurred an ongoing battle over religious freedom. Indigenous peoples from the Southwest and northwestern Mexico tribes have used peyote for ceremony and as a medicinal since antiquity. Peyotism involves

Christian sacramental ritual. The NAC, which can therefore trace it roots partially to Mexico, was spread by Parker and his adherents throughout the United States and Canada. According to Parker, the difference between the NAC and other religions is simple: "The White Man goes into his church and talks *about* Jesus. The Indian goes into his tipi and talks *with* Jesus."

Peyote and the NAC have been credited with saving the lives of thousands of American Indians who needed a path to help them gain a proper relationship with themselves, with their community, and with the spirit world. Members have expressed how the "medicine" has helped them to overcome alcoholism and drug abuse. In addition, as for the Navajo gentleman and his jar of peyote tea, the medicine has helped many native peoples deal with and even be healed of chronic and debilitating illnesses.

USES

Peyote is mostly used in structured ceremony and ritual. It is not taken by native peoples for recreation, as the cactus is considered a powerful being who must be treated respectfully. Some American Indians will carry a small piece of peyote in a medicine pouch around their neck or at the waist to protect against illness and sorcery. The cactus is made into an infusion and taken as a painkiller and to treat tuberculosis, rheumatism, arthritis, colds, fevers, and intestinal illness. A poultice is applied to wounds, cuts, and rheumatic and arthritic joints. The shamans of my people take peyote as a way to enter into other dimensions, with the goal of retrieving lost souls responsible for specific illnesses. We also take peyote during ceremonies meant to heal the people, the land, the plants, and the animals. Additionally, peyote is used for influenza, toothache, diabetes, snake and scorpion bites, skin diseases, blindness, and asthma.

IDENTIFICATION AND HARVEST

Unless you have been trained by an expert to find peyote, you would probably step right past it. Peyote is an extremely slow- and low-growing spineless cactus, rising only about 4 inches above the ground. Most of the growth is actually below surface level. The cactus is pale blue-green to yellow-green with reddish green shoots that resemble flattened spheres. It most often grows in crowded groups of mixed mature and young plants. Individual cacti are characterized by rounded humps separated by rib-like fissures. There are no spines, but small tufts of woolly yellowish hairs rise from several areoles. Flowers open during the day; they are usually pink to white or yellow, sometimes red. Peyote has been overharvested in its natural environment; this and poor harvesting techniques have convinced authorities in Texas to place this cactus on their Endangered Species List. In Mexico, where this native species is also threatened, it is now illegal to harvest any cacti: they are protected under CITES and the Mexican government. My people and other traditional practitioners of the peyote ceremony engage in an extensive purification and ritual prior to going on the "hunt" for peyote; nevertheless, as a traditional practitioner of peyote

Peyote is listed as endangered by several regional authorities, a victim of poor harvesting techniques and overharvesting.

ritual, I must advise against the harvesting of this revered being.

The most extensive and longest-running battle between native peoples' religious freedoms and the U.S. government has involved the NAC and its use of peyote, and laws regulating peyote use vary by state. Current U.S. federal law restricts the use of peyote to federally recognized tribes; however, many NAC parishioners are not part of federally recognized tribes. American Indians who are parishioners of the NAC are permitted to purchase peyote by licensed harvesters.

HEALTH BENEFITS

Peyote contains over 60 alkaloids, the predominant one (6 percent by weight) being mescaline, the compound behind peyote's psychoactive qualities. Mescaline and the other alkaloids act as natural antibiotics, antimicrobials, analgesics, antioxidants, anti-inflammatories, antidepressants, and anxiolytics. They have also been used to treat neurodegenerative diseases and as cholinesterase inhibitors.

PIÑON
PINUS SPP.

Family: Pinaceae
Parts Used: leaves, nuts, bark, wood, gum
Season: year-round
Region: North America

I will never forget the time, during a community gathering in northern New Mexico, when an elder was expounding about the unusually humid fall air. He was enjoying being able to smell things so vibrantly. Normally the humidity in places like New Mexico is very low (15 to 20 percent). The sinuses sort of shrivel up, and scents do not carry very far. Perhaps this is why autumn in the Southwest is so special. It is the time of year when the people from various indigenous communities roast the annual harvests of green chilis, squash, ears of corn, and piñon nuts. Piñon trees are held in very high esteem among people native to the Southwest, the Great Basin region, and parts of the Sierra Nevada foothills of California. Some believe the tree played a role in their creation. The Havasupai, for example, who have occupied the bottom of the Grand Canyon for centuries, have an origin story where Creator saves First Woman by causing a large hollow piñon tree to fall, permitting her to crawl inside the tree trunk and survive a catastrophic flood. After the waters receded, raging rivers created the Grand Canyon, where First Woman gave birth to a male child. Later she conceived a female child near Havasu Falls; it was from this location that the people emerged.

In an Apache creation story, the Original Being creates all light, land, animals, wind, and a person named Girl-Without-Parents. During the original process of creation, all the water in the world was situated on only one side of the planet, causing the globe to continually roll around. Pigeon was sent to investigate this problem. He found all the water and realized that it was beginning to move in the direction of the people and animals. Pigeon returned to where Girl-Without-Parents and the Original Being were talking and informed them that all the water was headed their way; all the new creation was going to be wiped out. Moving quickly, Original Being created a tall piñon tree. Girl-Without-Parents collected sticky

pitch from the tree and began to form a huge hollow ball, using the tree as an inner framework for the ball. Girl-Without-Parents and all the insects and other animals gathered in the piñon pitch ball and rode out the flood.

USES

When most native peoples think of piñon trees, the first thing that comes to mind is a source of extremely tasty and nutritious nuts. There is really no way to accurately describe the wonderful taste of piñon nuts. It is a very unusual flavor. They can be eaten raw but are best when roasted. The taste of roasted piñon nuts is probably best described as a Southwest umami: a lingering, pleasant mix of sweet, savory, sour, and bitter, underscored by pine and sage. Piñon nuts are so highly prized by American Indians that they remain a source of trade. Even now, some indigenous families supplement their income by collecting piñon nuts.

Piñon nuts are eaten both raw and parched. The meats are removed from the cone, ground into a paste, and used as a gruel. The meats are also dried and ground into a flour to be made into cakes, breads, and puddings, and mixed into pies and soups. The Havasupai will sometimes parch the nuts and grind the meats into a paste for consumption. Some native peoples in the Southwest will add boughs of piñon needles to their pit roasts in order to improve the flavor of the meat that is being cooked. One of my favorite snacks is dried venison meat mixed with partially ground piñon nuts and currants.

The piñon is so important to native peoples that it figures in the creation stories of several tribes.

Piñon nuts, considered a delicacy by many native peoples, and a spray of western red cedar.

Navajo hogans were most often made of piñon logs. This one is in Datil, New Mexico.

As a medicine, the needles of piñon are burned, and the resulting smoke is inhaled for colds. A poultice of the melted gum is applied to cuts, sores, wounds, and burns. A decoction of the inner bark of two-needle piñon is taken as an expectorant tea; an infusion of its needles is used as an emetic. A decoction of the needles is also taken for headache and colds, coughs, and to reduce fevers. The Zuni will dry the gum and grind it into a powder, which is then sprinkled on wounds as an antiseptic. Piñon gum is often chewed as a snack. It can also be warmed and used as a general adhesive. The pitch of two-needle piñon is used to make red dye and in basketry. The Goshute use a decoction of the gum of single-leaf piñon for worms or other intestinal parasites, and its cooked pitch is taken by women to stop menstruation and as a form of birth control. The Paiute take a decoction of piñon resin for diarrhea and rheumatism.

The Apache use the wood of young two-needle piñon for the main hoop of cradle-boards. Piñon nuts are also an essential food during the puberty ceremony for Apache girls. Piñon trees are prized for fragrant firewood and as a building material. Piñon logs are the central building material for Navajo hogans because the wood is resistant to rot and beetles. The bark of single-leaf piñon is still often used as a roofing material. Piñon resin is used by many basket makers to waterproof woven water jugs.

IDENTIFICATION AND HARVEST

Several species of piñon occur in North America, including some Mexican endemics. In the Southwest, the two main species are two-needle piñon (*Pinus edulis*) and single-leaf piñon (*P. monophylla*). The needles of piñon trees grow in bundles. The blue-green curved needles of two-needle piñon (which come in bundles of two) are short, growing only to about an inch in length. Two-needle piñon also tends to be a smaller tree or large shrub, 15 to 40 feet tall, but more often toward the shorter end. The thin bark is reddish brown with irregular furrows. The broad cones are small, rarely growing longer than the needles. Between the scales of the cone are found the treasured light brown nuts.

Single-leaf piñon tends to make a taller tree than its Southwest cousin. It regularly grows to approximately 35 feet. During its earlier years the tree is often rounded, sometimes pyramidal. As the tree matures, it will gain irregular branching and lose its rounded appearance. The bluish gray needles are cylindrical and usually one per bundle; they grow 1 to 2 inches long. The subglobose female cones are wind-pollinated; they ripen in August of the second growing season, full of half-inch-long, wingless seeds. Piñon trees produce cones every three to seven years. Trees usually do not start bearing cones before they are 30 years old and do not start producing good seed crops until they are about a century old.

Two-needle piñon is native to the high desert and highlands of the Southwest and Mountain West, including Wyoming, Utah, Colorado, Arizona, and New Mexico. It can sometimes be found in Oklahoma and western Texas. It is rare in the eastern parts of California. Single-leaf piñon is found in the piñon-juniper woodlands of the Southwest and among the Jeffrey pine forests and sagebrush steppe; its range includes the high Sierra Nevadas and the Peninsular and Transverse Ranges of southern and eastern California, into Arizona, New Mexico, and Utah.

Piñon gathering is an important and serious affair for many American Indian and Hispano families in the Southwest. The nuts are gathered during the fall, when they are ready to fall (or are already falling) from the cones. Many people will gather the fallen nuts from the ground or spread tarps and blankets around the base of selected trees, which are then shaken and lightly struck. Near-ripe cones are sometimes collected and then heated over coals or a fire to force them to open and release the nuts. Make inquiries in the local area prior to harvesting: you do not want to infringe on a stand of piñon that some indigenous or Hispano families might consider their property.

HEALTH BENEFITS

Piñon nuts are a good source of polyunsaturated and monounsaturated fats; they are 60 percent by weight linoleate, oleate, palmitate, and stearate oils. One hundred grams of piñon nuts provides about 30 grams of protein as well as many important minerals and vitamins. In addition, studies demonstrate that the bark, needles, and resins of *Pinus* species are a rich source terpenoids, flavonoids, tannins, and xanthones, all proven antioxidant, antibacterial, analgesic, antimicrobial, and anti-inflammatory agents.

PRICKLY PEAR
OPUNTIA SPP.

Family: Cactaceae
Parts Used: pads, fruit, roots
Season: spring, summer, fall
Region: North America

The Aztecs were one of the largest indigenous empires of the western hemisphere. True, they subjugated nearby native communities, but they also created systems of mathematics, architecture, and agriculture that outmatched in sophistication what was happening simultaneously in Europe and other places. None of this would have come to pass, however, if not for the prickly pear. The Aztecs did not originally hail from what is now Mexico City; according to legend, they originated far to the north, in a land they knew as Aztlán, believed to be our American Southwest. The legend suggests that for centuries the Aztecs were in constant migration, being pushed around and forced away from places they were trying to settle. This continued until a certain Aztec leader had a vision that prophesied that they should migrate south until they reached a place near a lake and saw an eagle holding a snake in its beak while perched on a prickly pear cactus. Wherever they beheld this sight, they were to settle. This place eventually became the Aztec capital of Tenochtitlan, later Mexico City. The chief components of the legend—snake, eagle, prickly pear—are displayed centrally on the Mexico's national flag. It is not only the Aztecs or present-day Mexicans who

A prickly pear cactus with ripe fruit.

hold the prickly pear in high esteem, however. It is considered a gift from the land by many American Indians and is widely used for both food and medicine.

USES

Prickly pear is a very nutritional food. I recall watching my grandfather sup on one of his favorite soups of chopped cactus, pinto beans, broth, and wild rice. My grandmother made this dish for him all the time. Meals such as

The national flag of Mexico depicts the legendary signals—snake, eagle, prickly pear—by which the Aztecs knew to settle there.

Young prickly pear pads are both food and medicine.

this may have played a role in his reaching the age of 106. I enjoy carefully collecting the young prickly pear pads, chopping them up, and adding them to soups and stews, or sautéing them with scrambled eggs. A Yaqui friend of mine enjoyed bean burritos with chopped prickly pear. Prickly pear can also be harvested and dried for later use. Prickly pear fruit (*tuna* in Spanish) is a delicacy around the Southwest. They can be eaten directly, but the many seeds inside the fruit often ruin enjoyment of the sweet juicy flesh. More often, the seeds are sifted from the fruits, which are then made into syrups and jams; the fruits can also be baked. The Apache save the sifted seeds, parch them, and eat them as a snack or grind them into flour.

For medicinal use, the pads of prickly pear are scorched and then split and used to stop the bleeding of a wound and to treat infections, cuts, and burns. The liquid from the boiled roots is taken to treat diarrhea and as a laxative. The warm fruit juice is rubbed onto the joints to treat rheumatism. An infusion of the smashed pads is taken to ease childbirth. In a pinch, the thick pads can even be scraped of their spines and used as a temporary splint for a broken limb.

IDENTIFICATION AND HARVEST

Prickly pears are evergreen perennials. Some species (e.g., plains prickly pear, *Opuntia polyacantha*) grow only 6 inches tall; *O. ficus-indica* and others grow up to 16 feet. They are hardy plants and not frost tender. The plants tend to spread in all directions and flower from June to September; later in the year the plants will produce reddish purple fleshy fruits that grow up to about 3 inches long. The pads are green, sometimes blue-green, generally obovate to roundish. Most prickly pear flowers are yellow, occasionally reddish; they are up to 3 inches wide and nearly as deep; the blooms open about mid-morning and will last an entire day.

Although prickly pears are icons of the Southwest and Mexico, the genus is widespread in North America; for example, *Opuntia humifusa* is native to New York, and *O. cochenillifera* can be found in Florida. When Lewis and Clark began their journey to explore the western regions of the continent in 1803, they expected to encounter wild animals, Indians, and perhaps challenging weather as they passed through what are now the Dakotas and Montana. What they were not braced for were mosquitoes, clouds of gnats, and prickly pear. The plains prickly pear posed a major difficulty, as its stout spines made easy work of the moccasins worn by many members of the Corps of Discovery. On 8 September 1805, William Clark lamented in his journal, "[O]n the head of Clark's [=Bitterroot] River I observe great quantities of a peculiar sort of prickly pear growing in clusters, oval and about the size of a pigeon's egg, with strong thorns which are so bearded as to draw the pear from the cluster after penetrating our feet."

Generally prickly pear does well in light sandy soils and does not do well in shade. When harvesting prickly pear for food, it is best to collect the young pads during the spring.

Older pads that have grown over 7 inches long begin to gather more bitter gallic acids and thicker fibers, making consumption unpleasant. After collecting, the glochids—the small nodules holding the spines—can be scraped off with a sharp knife. Harvest the fruits in late summer and fall. Be careful, as the fruits are also covered with many glochids that have tiny spines.

HEALTH BENEFITS

Opuntia species are significant sources of soluble fiber, mucilage, amino acids, minerals, carotenoids, antioxidants, polyphenols, quercetin and other flavonoids, and gallic acid. Prickly pears are valued anti-inflammatories and diuretics; they also prevent blood clots and help lower blood pressure.

RABBITBRUSH
CHRYSOTHAMNUS SPP.,
ERICAMERIA SPP.

Family: Asteraceae
Parts Used: whole plant
Season: year-round
Region: western North America

Rabbitbrush (with species in two genera, *Chrysothamnus* and now *Ericameria*) is ubiquitous throughout the western portion of our continent, particularly the Mountain West. Rabbits and other small animals use the shrubs for cover from birds of prey and other hunters. In the Basin and Range, sage grouse rely on rabbitbrush as cover for nesting. In winter and early spring, jackrabbits forage on the foliage. People too rely on rabbitbrush, for various cultural uses and as a medicine.

USES

The Hopi revere yellow rabbitbrush (*Chrysothamnus viscidiflorus*). It is one of the plants that they add to their prayer sticks as a decoration. The Hopi will also take the woody stems of rabbitbrush and use them to build wind- and sandbreaks for their young corn plants. They also encourage the growth of the wild shrubs at the edges of their corn fields. Rabbitbrush blossoms are used as a yellow dye by Navajo weavers.

Rabbitbrush is used by several western tribes as a medicinal. The Paiute make an

A contemporary Navajo Teec Nos Pos rug by Glenna Begay, created using traditional techniques and materials, including rabbitbrush dye, the source of the gold field here.

infusion of the foliage to treat colds and coughs. They will also line the floor of sweat bath areas with rabbitbrush in order to treat rheumatism. The Hopi use its leaves as a poultice to treat boils. At Isleta Pueblo the whole plant is made into a decoction and taken to reduce fevers. The stems are chewed to relive toothaches. An infusion of the plant is taken as a cold remedy and for sore throats. The Navajo also drink a strong infusion as a ceremonial emetic.

Rubber rabbitbrush (*Ericameria nauseosa*) is used similarly to *Chrysothamnus* species; additionally, the Navajo will take a decoction of the root for menstrual pains and the flowers for stomachaches.

IDENTIFICATION AND HARVEST

Most species of rabbitbrush are low-growing perennial shrubs, topping out at about 3 feet.

Rabbits use this shrub to hide from predators, giving it its common name.

However, in the right conditions, yellow rabbitbrush grows up to 5 feet, and rubber rabbitbrush can reach 8 feet. The leaves of rabbitbrush are typically narrow and linear, growing 2 to 3 inches long. They often appear a bit twisted and sometimes seemingly folding up on themselves. The leaves are also resinous and sticky, leading to the mistakenly applied common name, greasewood. The leaves are borne on woody stems that are often covered with tiny hairs; bark is pale green to grayish. Beginning at the end of summer and into the fall, bushy terminal flower heads top the stems of rabbitbrush; the individual flowers of most species of rabbitbrush are one of several shades of yellow to yellow-orange, and in some species (e.g., *Ericameria nauseosa*), the flowers have a pungent smell. The fruit that is produced at the tips of the stems is actually a tiny hairy achene, ending with several fine feathery bristles that help the single seed to float away in the breeze when released. Look for rabbitbrush in sagebrush and woodland habitats throughout the western and drier regions of North America. It does well in alkaline, calcified, and saline soils and quickly establishes itself in disturbed habitats, such as burns, flooded arroyos, and rockslides.

HEALTH BENEFITS

Chrysothamnus species contain several pharmacologically active compounds, including sesquiterpene lactones, cinnamic acids, and coumarin glucosides. Sesquiterpene lactones are anti-inflammatory, analgesic, and antimalarial. Cinnamic acids are antimicrobial.

Coumarin glucosides demonstrate anticancer, antitumor, antibacterial, and anticoagulant activity.

RABBIT TOBACCO
PSEUDOGNAPHALIUM OBTUSIFOLIUM

SWEET EVERLASTING, FRAGRANT EVERLASTING, WHITE BALSAM

Family: Asteraceae
Parts Used: whole plant
Season: summer, fall
Region: eastern North America

Trickster figures are abundant in indigenous stories. They are inherently selfish, and their actions are often a result of thinking only about their own wants and needs. However, through their mischief and antics, they often cause good things to happen to everyone around them. The best-known Trickster figure in North America is Coyote, but among many eastern tribes it is Rabbit. Rabbit saved the Penobscot from starvation, brought the moon to the sky for the Oneida, and defeated a man-eating monster on behalf of the Alabama. In one story, while Rabbit is engaged in one of his antics, he gets tangled up in a thicket; in the process of detangling himself, he encounters a sweet-smelling plant, which he uses afterward to heal the cuts he got from the thicket. In another story Rabbit smokes a special plant in order to communicate with the Great Spirit. Today, native peoples add rabbit tobacco to smoking mixtures and use it to treat cuts.

Sweet-smelling rabbit tobacco is often included in smoking mixes.

identifying characteristic of rabbit tobacco is its scent. To some, it is suggestive of a mild tobacco; others describe the fragrance as being reminiscent of vanilla or maple syrup. In any case, the distinctive scent stays with the plant even after it dries up during the late fall and over winter. Rabbit tobacco is typically found in open meadows and fields and does well in cultivated or disturbed areas. It is an eastern plant, distributed from the Canadian Maritimes down to Louisiana and west to the

USES

An infusion of the whole plant is used to treat scratches and cuts as well as muscular cramps and rheumatism. A decoction of the plant is used as a face wash to treat insomnia and nervousness and taken to treat colds, cough, and lung pain. A decoction of only the leaves is taken to treat vomiting, fevers, and coughs. The dried leaves are also used as a smudge for headache and fainting and are added to ceremonial smoking mixtures.

IDENTIFICATION AND HARVEST

Rabbit tobacco is a biennial herb. In its first year, in most regions, the plant produces a tight, woolly-haired basal rosette of leaves. In its second year, the plant produce a tall stem that grows up to about 3 feet tall. The stem leaves are simple and alternate, and the flowers are actually clusters of tight, white, peg-shaped discs with yellow centers. A key

A sweet scent of vanilla or maple syrup can help identify this often-overlooked herb.

Mississippi River. Many gatherers collect the entire plant during the fall and use the dried leaves and flowers over the winter.

Plants display some 125 different phytochemicals, including flavonoids, sesquiterpenes, diterpenes, triterpenes, phytosterols, anthraquinones, and caffeoylquinic acid. These constituents have various pharmacological applications, being, among other things, antioxidant, antibacterial, antifungal, antitussive, anti-inflammatory, antidiabetic, and antihyperuricemic.

RED OSIER DOGWOOD
CORNUS SERICEA

Family: Cornaceae
Parts Used: leaves, branches, fruit, bark
Season: year-round
Region: North America

American Indians who practice ancestral ceremonials are often asked the same question: what is it, exactly, that we smoke in our ceremonial pipes? Most indigenous peoples in northern and western North America use a mixture of leaves and bark from several plants in our smoking mixtures. In the Northeast, the mixture often includes the inner bark of red osier dogwood, as well as bearberry, tobacco, and various bunchberries (*Cornus* spp.) and sumac (*Rhus* spp.). But red osier dogwood is more than a smoking ingredient. It is also used by American Indians as a medicine and for basket making.

USES

Besides being an important ingredient in several indigenous smoking mixtures, the bark of red osier dogwood (and its ssp. *occidentalis*, western dogwood) is used by many tribes in the form of an infusion to treat sore eyes, snow blindness, and colds. An infusion of the bark is applied to wounds to stop bleeding and to various parts of the body as an analgesic. It can be gargled for sore throats and drunk to treat diarrhea and as a general tonic. An infusion of the bark is consumed in large amounts as a ceremonial emetic. A watered-down decoction of the bark is consumed to treat fevers. In addition, a decoction of the twigs and leaves of dogwood is used to help a woman heal after childbirth. Blackfoot Indians have used the juice from red osier dogwood berries as an

The inner bark of red osier dogwood, here in bloom, is a classic ingredient in ceremonial smoking mixtures.

White berries form in late summer to early fall.

arrow poison to cause infection in victims. This is a curious use, as the berries of red osier dogwood can also be mixed with serviceberries (*Amelanchier* spp.) as a tart snack. The branches and twigs of red osier dogwood are very pliable, making it a useful basket-making material. If the branches are gathered during late winter or early spring, the bark will maintain its red color. Some native peoples use the bark for tanning hides. The straight young shoots are used for making arrows.

IDENTIFICATION AND HARVEST

Red osier dogwood is a woody deciduous shrub, usually 4.5 to 10 feet tall, occasionally to 20 feet. The shrubs have a roundish arching shape with many twigs and branches. Twigs are red-brown in autumn, becoming bright red during the winter months; slice open a good-sized branch, and you'll find a whitish pith running down the center. The red-brown bark becomes dark green during the summer. The 2- to 4-inch leaves rest opposite each other

The vivid branches of red osier dogwood lend color to winter landscapes across North America.

on the branches. The leaves are ovate and smooth-edged, and tend to be lighter green and hairy underneath, medium green above; they turn reddish purple in fall. In summer, the shrub produces umbrella-shaped clusters, 1.5 to 2 inches across, of white flowers; by late summer, clusters of white berries begin to form. The small berries will have a dark stipple on their ends and furrows on the sides. Look for red osier dogwood throughout

North America in moist soils along streams and creeks and in open forested areas. The shrub responds well to coppicing and pruning. During the late fall, native basket makers often coppice the shrub down to about 6 inches above the ground in order to promote straight shoots the following spring.

HEALTH BENEFITS

Red osier dogwood bark is rich in healthful tannins, and berries and other plant parts contain phenols and flavonoids. One study showed an increase in concentrations of the antioxidant flavonoids (rutin, quercetin, anthocyanin) in spring and summer and a decrease in the fall.

RED ROOT
CEANOTHUS SPP.

Family: Rhamnaceae
Parts Used: leaves, flowers, fruit, roots
Season: year-round
Region: Southwest, Mountain West

Pueblo Indian and other native communities maintain a year-round ceremonial cycle. The clan singers among the Pueblo and spiritual singers among the Navajo and Apache, for example, must be prepared to sing nonstop, sometimes for hours. Navajo Blessingway ceremonies last three to nine days. It is not really considered a "ceremonial" or sacred plant, but red root does play an important role in helping the ceremonial chanters and singers maintain

Red root plays an important role in helping ceremonial singers maintain their vocal cords over several days of duties.

their vocal cords during these important events. Red root, however, is more than a throat gargle; it has several medicinal and non-medicinal uses.

USES

Red root is an anticoagulant; therefore, it should not be used in combination with modern anticoagulant or blood-thinning medications. However, it is the anticoagulant property of red root that makes it useful for situations where thinner blood might be called for, such as with high cholesterol or headaches after fatty meals. Red root is infused as a gargle and used by native peoples for sore throat, tonsillitis, and pharyngitis; in addition, a decoction of the root is useful as a general tonic for thick, viscous blood. The flowers of red root have been made into a hair tonic; the leaves of the plant make a mild tea. During periods of famine, Navajo and Pueblo peoples would eat the little berries that grow on the plant.

IDENTIFICATION AND HARVEST

In the West, five *Ceanothus* species are red roots, but in this entry I will focus on just two. *Ceanothus greggii* grows in thickets, sometimes forming a ground cover, sometimes reaching 5 feet tall; its bark is nearly white, whereas that of *C. fendleri* approaches claret-red. The little oblong leaves are thick on the tough, spiny branches, held alternately in groups of two or three. They are grayish green and very thickly hairy underneath; sometimes they emit a wintergreen scent. The white to sometimes pinkish blooms come all at once in June and July, although they can bloom as early as April and linger as late as October. The roots of both species are tough, deep, and huge for what are

The color of their roots makes it clear how these robust shrubs got their name.

normally smallish shrubs. Red root can be collected any time of the year, although native herbalists look for the plants that appear to have been more environmentally stressed than the others: it is believed that stresses on the plant cause it to produce more of its medicinal compounds. When collecting the root, be prepared to dig into hard-packed rocky soils and around other spiny shrubs. The root is very tough and difficult to cut up into smaller pieces. The leaves and flowers should also be collected with care as the branches are spiny.

HEALTH BENEFITS

The anticoagulant properties of *Ceanothus* species are attributable, in part, to the effects of two kinds of acids, including four dicarboxylic and two phosphoric acids. Peptide alkaloids and antimicrobial compounds (including three triterpenes, two flavonoids, and ceanothus acid) have also been identified in the genus; the ceanothus acid is especially effective against streptococcus.

SAGEBRUSH

ARTEMISIA SPP.

Family: Asteraceae
Parts Used: whole plant
Season: year-round
Region: western North America

Some plants are both synonymous with American Indian herbology and icons of a particular geographic region. Sagebrush is one of those plants. For people who study the indigenous knowledge of plants of the Greater Southwest as well as the entire region west of the Mississippi, sagebrush is always at the forefront. It grows nearly everywhere in the region, which accounts for its status as a marker of the romanticized West. Every time I catch the unmistakable resinous odor of sagebrush, a rolling release of neuropeptides fills my head, activating scores of cultural and personal memories related to ceremony, ritual, the spirit world, visions, ancestral knowledge, and family. Attend any American Indian event even

Sagebrush is one of the iconic plants of American Indian traditions, valued for its scent as well as its many medicinal properties.

slightly connected to traditions or culture, and sagebrush incense will be burning or used as a smudge to purify the space. It has such an incredible amount of medicinal and other uses that it should be a central element of any plant specialist pharmacopeia.

USES

My people use *rosawari*, as we call it, for diarrhea and general stomach problems, menstrual pains—and for just about everything else, it seems. Artemisias in the Southwest and West are used by American Indians in many similar manners. Sagebrush is a general analgesic which is applied as a poultice of leaves for toothache, cramps, and painful menses and made into a bath for rheumatism sores. It has been made into a poultice and infusion to treat wounds, as a burn dressing, and for eczema. An infusion of the roots can be applied to the scalp for skin infections. A tea or decoction of the leaves is used as an antidiarrheal, for fever, and to treat colds, coughs, and intestinal and other kinds of parasites. Many native peoples use the dried leaves as an incense or ceremonial fumigant. I cannot recall any ceremony or native event that did not begin with a ritual smudging of sagebrush. Navajo weavers boil the leaves and twigs in order to create different shades of yellow and gold dye. Here are some more specific uses, by species.

A decoction of *Artemisia dracunculus* is used for eyes, colic, and headaches; a poultice is used for headaches and diaper rash.

Artemisia frigida is made into a tea that is used as a pulmonary aid, an abortifacient, and a diuretic, as well as for heartburn, convulsions, skin irritation, cough, flu, and tuberculosis. As a poultice, it is used for ceremony and to treat bleeding wounds.

A poultice of *Artemisia ludoviciana* is used as a foot deodorant as well as for sprains and swellings, sore feet, and rashes. A tea of this species is good for taking away bad dreams and reducing phlegm; it treats the flu, sore throat, sinus issues, and constipation. Some will take a sweat bath using the leaves to steam out infection and flu and for protection against bad energies and broken taboos.

An infusion of *Artemisia vulgaris* is good for cold, flu, and sinus problems, and as an expectorant.

IDENTIFICATION AND HARVEST

Artemisia is a large genus in the daisy family, approaching 400 species by some counts. Depending on the species, plants are 2 to 7 feet tall. Some species (e.g., *A. ludoviciana*) have lanced leaves; most have irregularly lobed leaves, up to 2 inches long. *Artemisia tridentata*, as its epithet suggests, has regularly tri-lobed leaves. The surfaces of the leaves are typically covered with short fine hairs and are normally silver-green or gray-green, although some species can be nearly plain green. If I am unsure of what I am looking at, I will pinch a small piece of one of the leaves, crush it between my fingers, and sniff for that resinous scent. Very small clusters of yellowish white composite flower heads spring up from leaf axils in summer. In the Southwest, plants can be found in large dense stands at the base of rocky outcrops and growing in smaller stands on hillsides and mountains up to 11,000 feet in elevation. All parts of the plant are useful and can be easily collected with scissors or clippers. Before tossing a bag of your collections in the car with you, be mindful that plants are very aromatic. Sagebrush can be collected any time of the year.

HEALTH BENEFITS

Artemisia species contain numerous essential oils, which accounts for the characteristic sagebrush aroma; one study found 74 different essential oils among several species, including thujone, an emmenagogue, and other pharmacologically effective compounds. The genus also contains lactone glycosides, which may account for sagebrush's anthelmintic and diaphoretic properties.

SAGUARO
CARNEGIEA GIGANTEA

Family: Cactaceae
Parts Used: fruit, ribs
Season: summer, fall
Region: Southwest

Every summer, normally sometime in July, Tohono O'odham and Pima people venture into the hot Sonoran Desert armed with very long sticks that have a shorter stick attached perpendicularly at the end. The sticks resemble a 10- to 15-foot-long crucifix. This time of year is known to the Tohono O'odham as Hashañ

Mashad (Saguaro Harvest Moon). They will gather the sweet ripe fruits that sprout from the tops of the tall cacti in preparation for the Vikita: this harvest ceremony is a crucial time for the Tohono O'odham and Pima as it ensures that the summer monsoon rains will come and nourish their crops as well as the sensitive flora and fauna of the desert.

Tribes that call the Sonoran Desert home believe that the saguaro was once human. One legend suggests that a little boy fell into a hole in the ground, and when he finally emerged, he was a saguaro. Another legend describes how a mischievous little boy and his brother were arguing and accidentally knocked over a clay jar full of precious water. In order to avoid getting into trouble with their grandmother, they escaped into the desert. Once there, the two brothers decided to transform into desert plants. One boy changed into a palo verde tree; the other became a saguaro so he would live forever and bear edible fruit for the people. The Tohono O'odham believe that every saguaro in the desert represents the place where a bead

Long picking poles used in the annual saguaro fruit harvest.

Heart in hand: ripe saguaro fruit from Saguaro National Park.

of sweat fell onto the ground from the brow of their Trickster creator, I'itoi. Saguaro is not only a keystone Sonoran Desert species (i.e., its role is central to the functioning of local ecosystems), it is also central to O'odham and Pima culture.

USES

To American Indians, saguaro is mainly a source of sweet food; it is also an important ceremonial plant and provides materials for tools and construction. The sweet fruits that grow atop the columnar cacti are collected and eaten directly. The insides of the fruit are scooped out and made into syrups and preserves. The sweet juice is mixed with other foods as a sweetener. The juice is allowed to ferment into an alcoholic drink that is consumed during Tohono O'odham rain house ceremonies. Saguaro fruit juice can also be filtered

A Tohono O'odham "boot" canteen, circa 1915, made from the casing around an abandoned bird nest bored into the side of the saguaro; the plant forms watertight scar tissue on the interior of the cavity.

and taken as a refreshing beverage. The fruits may also be dried for later use, in which case the seeds of the fruit are separated from the pulp, allowed to dry or are parched, and later ground into flour that is used for making cakes, breads, and porridge.

Saguaro cacti are very long-lived; many live for over 400 years. When a saguaro does finally die, its inner ribs are exposed. The medium-hard ribs are a source for the long picking poles used in the annual saguaro fruit harvest; they are also used to make cooking implements and for fencing, roofing houses, and as warp material for basket making. The ribs can be split and made into small traps, cages, and rough baskets, and many other tools and construction materials.

An interesting symbiotic relationship occurs among saguaro, gila woodpeckers, and several other desert birds. The woodpecker drills holes into the sides of the cactus in order to get to the moisture inside and then nests in one of the resulting cavities. Eventually, scar tissue from the injury to the cactus develops, creating a thick layer around the entire interior of the cavity. Later, when the nest has been abandoned by the woodpeckers, other desert birds—screech owls, elf owls, flickers, and many others—will occupy the hole. Finally, after the cactus has died, desert dwellers such as the Tohono O'odham, Apache, or Pima collect the hard-skinned burls, or "boots," to be used as containers for water and other items.

IDENTIFICATION AND HARVEST

In the Sonoran Desert, look for the tallest columnar cactus in the vicinity, whether growing nearby or in the distance, and you have found saguaro. This cactus is characterized by an erect central stem. During its younger years, the central stem grows to about 10 feet tall and 2 feet wide. At maturity, after about 100 years, the cactus can reach heights of up to 40 feet with a trunk diameter of around 30 inches;

A saguaro will not begin to sprout arms until it has reached its full mature height.

this is when the cactus begins to sprout arms. Saguaro is also characterized by its prominent ribs, 12 to 30 in number. The stem and arms are covered with dense spines. Throughout their lifetime, saguaros sprout funnel-formed flowers from the tops of the central stems and arms; the flowers are white, with a creamy yellow central boss. From the flowers emerge the scaly dark red fruits, which grow to about 2 inches long.

White saguaro blossoms precede the fruits.

Saguaro fruits are an excellent source of important minerals and vitamins C and B12; they are also high in soluble and insoluble fiber. Although saguaro is not really thought of as a medicinal plant, Apache, O'odham, and other desert peoples feel strongly that eating saguaro fruits helps to prevent arthritis and rheumatism.

Saguaros are endemic to the Sonoran Desert. Their northernmost occurrence is about 30 miles north of Phoenix, Arizona; they are richest in the Tucson area and grow south into the northern half of Sonora, Mexico. They are sometimes found near the California–Arizona border but quickly dissipate as one travels east of Phoenix or Tucson or goes above 2,500 feet in elevation. The cactus prefers gravelly slopes and rocky ridges.

Harvesting saguaro fruit is difficult. Like the people indigenous to the Sonoran Desert, you will need some kind of long pole, tall enough to reach the tops of the central stem and arms. The pole should extend at least 20 feet and have some kind of hook or perpendicular arm for pulling or knocking off the fruit. To prepare the fruit for eating, cut the fruit open, scrape the pulp into a vessel, and strain the seeds from the flesh. In order to eat the seeds, let the pulp dry and then pound it to separate the seeds. Approximately 25 pints of fruit make a gallon of syrup.

SAW PALMETTO
SERENOA REPENS

Family: Arecaceae
Parts Used: whole plant
Season: year-round
Region: Southeast

In many parts of the Southeast, saw palmetto is a keystone species and central to the culture of indigenous communities, providing essential materials for basketry, construction, medicine, and ceremony.

USES

The Choctaw use the stems in basket making. The Seminole separate the fiber from the large fan-like leaves and stems and use it for making baskets, cordage, rope, and brushes. The ripened red fruits are eaten, as are the young tender shoots. The fruit can be dried for later consumption and also crushed for its juices. Extract of the fruits of saw palmetto is used as an expectorant, as a general tonic, as a tonic

The fibrous leaves of saw palmetto make it useful for baskets and cordage in particular.

Seminoles shape palmetto fiber to create the body of children's dolls.

for prostrate problems, as an antiseptic, and to treat impotence and inflammation. The fruits have also been used to treat infertility in women, to increase lactation, and to decrease painful menstruation cycles. The large leaves are used to make ceremonial dance fans, hats, and roof thatching, and the trunk of saw palmetto makes a good punk material for starting fires. Many parts of the plant are made into children's toys; the dried leaves are bunched together for bedding material.

IDENTIFICATION AND HARVEST

Saw palmetto is a low-growing shrub-like palm found from Florida west to Louisiana and north to South Carolina. The stems grow horizontally and are often branched. The fan-shaped leaves grow up to 3 feet across and are divided into 20 to 30 segments. The petioles are up to 18 inches in length and edged with curved teeth. The plant produces up to five panicles of small, cream-colored, fragrant flowers. The unripe fruits resemble large green olives; they turn bluish, reddish, and finally black as they mature, from midsummer into October. Saw palmetto often grows in large unbroken stands extending for miles. The sharp teeth of the petiole will break the skin, so care must be taken when collecting the leaves and fruits.

HEALTH BENEFITS

Saw palmetto fruits contain several fatty acids, including lauric, capric, caprylic, myristic, and

Unripe saw palmetto fruits resemble olives; they turn black as they ripen.

palmitic. A lipodosterolic extract of the fatty acids demonstrates anti-inflammatory, androgenic, and potential anticancer effects.

SLIPPERY ELM
ULMUS RUBRA

Family: Ulmaceae
Parts Used: branches, bark, wood
Season: spring, summer, fall
Region: eastern North America

When most people think of slippery elm, I imagine that the first thing that comes to mind is those little pink tins of cherry-flavored throat lozenges taken for sore throats and hoarseness. What many folks may not realize is that slippery elm is a large tree that offers so much more than a soothing throat balm.

USES

Among the Cherokee, slippery elm bark is made into a poultice and applied to help heal sore eyes, wounds, sores, and burns. The bark is used for floor matting when camping in snow. The Menominee boil the bark and use it to weave baskets, bedding, and mats. The bark is also woven into frames for snowshoes and fishing nets. Early in the 18th century, French missionary Joseph-François Lafitau reported that a group of Iroquois constructed a coarse canoe from a single large sheet of slippery elm bark, using branches from the same tree for the framing of the temporary craft. Plains Kiowa used slippery elm branches for the frame of their horse saddles. The inner bark of slippery elm is made into a decoction or chewed by many eastern native peoples to treat sore throats and coughs. A decoction of the inner bark is given to birthing mothers who are experiencing a prolonged labor. An infusion of the inner bark is used to soothe diarrhea; in

Slippery elm leaves help to identify the tree, which is valued for the medicinal properties of its bark.

smaller amounts, that same infusion can be used as a laxative.

IDENTIFICATION AND HARVEST

Slippery elm is a beautifully arching tree that can grow to about 60 feet. The thick, 4- to 7-inch leaves are alternate, obovate, and serrate, with obvious pinnate nonequilateral veins. The pinkish flowers with orange to yellow centers appear before the leaves in the spring; they sit very close to the twigs and branches in bundled clusters. The fruit is a greenish flat samara that encloses a single seed. The outer bark is gray and deeply fissured; the inner bark is mottled, fibrous, reddish yellow to reddish tan. The outer bark should be slippery or mucilaginous on the inside. Look for slippery elm trees in moist soils on lower slopes near streambanks and bottomlands. It will usually be growing among other deciduous trees such as boxelder, ash, walnut, silver maple, red oaks, and cottonwoods. Collecting slippery elm is a delicate process. During warmer weather,

A Kiowa saddle frame, circa 1909.

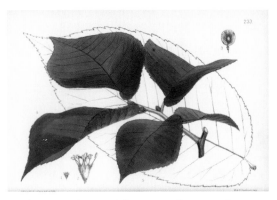

A 19th-century botanical illustration of slippery elm leaves, inflorescence, and flattish samara.

Slippery elm bark prepared for medicinal use.

the bark of slippery elm peels away from the tree trunk without too much work; during colder weather, it takes some careful effort that involves hot water. Try not to create too much damage to the tree when harvesting the soft inner bark. Carefully remove the bark from the tree by pulling off 4-inch-wide strips. Collect only a small amount of bark from a single tree; this will permit the bark to regrow.

The biochemical components of slippery elm include mucilage, tannins, flavonoids, and capric acid, a fatty acid with antibacterial, antifungal, and anti-inflammatory properties. The carbohydrates in mucilage swell when wet to form a viscous substance that can play a role in coating mucous membranes, helping to relieve coughing. Tannins are astringent, anti-inflammatory, and antioxidant; flavonoids too are known antioxidants.

SPANISH MOSS

TILLANDSIA USNEOIDES

OLD MAN'S BEARD

Family: Bromeliaceae
Parts Used: whole plant
Season: year-round
Region: Southeast

My grandfather used to tell me stories about plants that were each others' friends. For example, Drinking Gourd used to hang out with Tobacco and cause different kinds of mischief. Wásia (osha) was a friend of Sitakame (Mexican logwood). And Hikuli (peyote) is a companion to Dowaka (Spanish moss). I often wonder how some of these plant friends met each other. Wásia and Sitakame come from very different ecosystems; they could not possibly have met each other naturally. Similarly, Hikuli is confined to extremely hot and arid parts of the southern Chihuahuan Desert,

Spanish moss is in fact a flowering plant, not a moss.

whereas *Tillandsia* species grow in humid environments. I can only surmise that these are long-distance spiritual relationships.

USES

Native peoples, especially those from the Southeast, where Spanish moss grows naturally, collect, clean, and detangle the plant to reveal the fibers of the plant. It can then be woven into coarse blankets that are used for floor mats and horse blankets. The fibers are twisted into string, cordage, and rope. The dried fibers are used to remove scum from the inside of cooking utensils. Dry Spanish moss can be wrapped around a stick, lit on fire, and used as a temporary torch. The Seminole use Spanish moss as a tanning agent for deer skins. The moss is used as stuffing for toy dolls. Spanish moss is also a medicinal. An infusion of the plant is taken to treat chills and fever. It can also be taken as a purgative and a laxative, and as a wash to treat rheumatism.

Traditional baskets and a blanket woven from Spanish moss by members of the Alabama-Coushatta tribe.

IDENTIFICATION AND HARVEST

Spanish moss is a perennial epiphyte. Its fibrous blue-green stems grow suspended from the limbs of live oaks and other southern trees, sometimes as long as 30 feet from their origin point. In the South, Spanish moss is also found hanging from telephone lines and fences. A close inspection of the long, branching stems reveals that they are composed of overlapping scales, which work to collect water and nutrients from the air. Many small three-petaled pale green flowers are found at the axils of the leaves; between spring and fall, the flowers produce tiny sailed seeds that float on the wind from the plant. The plant has no roots. Spanish moss is not really Spanish in origin; it is native to the warmer humid climes of North America. It can be found on live oak and pines near swamps and rivers, and along the coastal areas of the Southeast. It is easy to collect the moss from where it is hanging. The process used for stripping off the outer scales and coating is a bit more involved; the plant must be soaked for at least six weeks to encourage the scales to rot away.

HEALTH BENEFITS

Tillandsia species contain the minor flavonoids retusin and artemetin, both of which are analgesic and anti-inflammatory.

SQUASH BLOSSOM
CUCURBITA SPP.

Family: Cucurbitaceae
Parts Used: male flowers
Season: summer, fall
Region: North America

I do not recall the first time I ate squash blossoms. I do remember attending a feast day at Ohkay Owingeh Pueblo, New Mexico. Accompanying the colorful dances was an abundance of local dishes. In the adobe home of an acquaintance was a dinner table covered with various offerings. I recognized pots of beans, a lamb stew, handmade oven breads, and a platter of what I assumed were fried chilis rellenos. I filled my plate with some beans, some bread, and the rellenos but after one bite realized that it was squash blossom stuffed with goat cheese, battered and baked, and then smothered with some of that local New Mexico red chili sauce that tastes like the very soul of the region.

Among Hopi cooks and other Pueblo Indian communities, the blossoms are most often chopped into small pieces and added to soups, bean sprout and lamb stews, and other dishes. Many cooks in Hispano and Pueblo kitchens will stuff the blossoms with cheese, chilis, beans, and spices to be baked. Other people in the Southwest like to batter the blossoms and deep fry them. The blossoms have a mild flavor and thus can be added to salads and other vegetable mixes without overtaking the featured flavors.

During the second half of the summer in the Southwest, both American Indian and Hispano families eagerly await the blossoms that create constellations of yellow stars in their fields of summer squash and pumpkins. The yellow flowers among the large green leaves are indicators of what the upcoming harvest of squash and pumpkins will be like. We often grew yellow summer squash (aka banana squash) when I was growing up. To this day when I raise summer squash, I can recall as a little boy being fascinated by the beetles that sometimes infested the plants; they resembled little gray armored tanks. It turns out that the beetles damaged the plants, threatening the coming squash harvest, and my grandmother encouraged me to pick them off and save her plants from destruction. Her healthy plants were eventually covered with yellow stars, and the best memory of all is how, after a round of beetle-picking, I would gather some of the male flowers so that Grandma could make me one of my favorite dishes: squash blossom quesadillas.

USES

For millennia, squash has been one of the most important foods in the Southwest, and the prevalence of squash blossom iconography, dating back to pre-Columbian art and artifacts, highlights the cultural significance of squash and squash blossoms to the indigenous peoples of the region. Potters from Tesuque, Hopi, and other Pueblos are noted for their use of stylized squash blossoms in the finished glazes; there is a Hopi katsina depicting a squash spirit; and squash blossom necklaces are famous among native southwestern

Squash blossoms can be stuffed with other ingredients or chopped and added to soups and stews.

A silver necklace featuring squash blossom iconography.

silversmiths: stylized blossoms are shaped from tubes of sterling silver and circle the neck of the wearer protectively.

IDENTIFICATION AND HARVEST

Squash and pumpkin plants will produce an abundance of large mustard-yellow star-shaped blossoms beginning about mid-summer and continuing into the early fall. Although both the female and male blossoms are edible, only the female blossoms, once pollinated, will mature into a pumpkin or squash; therefore, one must look for the male blossoms for harvest. Be careful not to over-harvest the male blossoms, however, as they are required for pollination of the female blossoms. The male blossoms can be identified by the tall stamens rising from the center of the flower; male blossoms also have longer stems than the females. The stems can be up to 7 inches long. Harvest of the male blossoms can begin once they reach the deep yellow color so typical of squash, zucchini, and pumpkin plants. They can be harvested by hand, but be sure to wear gloves, as the stems are often covered by hundreds of tiny prickly hairs. My grandmother used to send me out to where our squash was growing armed with a kitchen knife; these days, I collect squash blossoms with garden clippers.

HEALTH BENEFITS

Squash and pumpkin blossoms contain vitamins C, A, and a mix of beneficial Bs, including folate. They also contain iron, magnesium, potassium, selenium, phosphorus, and other nutritious minerals.

SQUASH AND PUMPKIN
CUCURBITA SPP.

Family: Cucurbitaceae
Parts Used: leaves, fruit
Season: summer, fall
Region: North America

Every fall, with keen anticipation, I watched my mother pour some olive oil into one of our large cast iron skillets in preparation of making calabacitas, one of my favorite seasonal dishes. After the oil heated up, she added to the hot oil the onions she had been mincing on our old chopping board. Next she added chopped garlic. She would allow that mixture to slowly cook while the scents and aromas wafted through our little kitchen; it seemed as if she drew this part out purposely, in order to whet my appetite. As the onions and garlic were

Calabacitas, a classic Southwest dish with a base of squash.

Cucurbita species are exclusively native to the Americas.

sautéing, she would mince some cilantro and Mexican oregano and toss that into the mix. Next she added the yellow summer squash, which she normally cut into narrow disks. The squash cooked for about a minute or two, and then my mom added some pinto beans that we had grown in our garden as well as some fresh sweet corn shucked from the cob. Now my appetite was really raging. But Mom had to add a few more ingredients before I could sate my hunger: some fresh chopped tomatoes, chopped jalapeño peppers, and finally some salt and black pepper. Calabacitas is only one of many dishes that can be made from the various squash and pumpkins grown by American Indians.

Squash and pumpkin are native to the Americas, originating in what is now Mexico around 10,000 years ago. These plants have since developed into 25 wild and cultivated species, diverse in fruit, color, size, and shape, and are now grown worldwide. One of the first to be cultivated, *Cucurbita pepo*, has formed the basis for most of the squash and pumpkins grown by native peoples from Mexico up to Canada.

The word "squash" is derived from the Narragansett word *askutasquash*. My people, the Rarámuri, use the words *bachi-ki* and *sito-bachi* to describe pumpkins and squash, respectively; we call pumpkin seeds *bachi-raka*. The Hopi call pumpkins and squash *patung* and pumpkin seeds *patangsivosi*; they refer to dried squash as *saaviki*. The Navajo call pumpkin *naayiziztossoi* and squash *naayizi*. This preamble would be very long if I were to list the names from each tribe in North America, but the point is that these cucurbits have always been an important food source.

USES

Today we refer to squash and pumpkins with terms that describe their shape, origin, and whether or not it is a summer or winter squash. This designation is important, as it determines how and when the fruits are grown, collected, stored, and eaten. The softer-shelled summer squashes are normally sliced and then fried, sautéed, boiled, roasted, or baked. They are added to soups and stews. The thicker and harder-shelled winter squashes and pumpkins are often stored in a cool dry place for later use. In the Southwest, Pueblo people store the harvest in the back corners of their adobe structures. The Navajo will dig a nearby pit lined with stones, where they will store their harvest of pumpkins and other vegetables. Winter squash is cut in half so that the seeds can easily be removed for later use. The soft inner flesh is baked, boiled, or roasted. It can be eaten directly or puréed into a base for sauces or soups. Sometimes the whole or halved pumpkins are baked on coals in open pits or slow-baked in closed pits. Many native peoples will cut the fruits into circles so that they can be dried in the sun on racks.

The seeds of winter squash are eaten throughout the Americas. They are roasted and eaten for snacks. They are also baked, hulled, and ground into a kind of marzipan. Sometimes the paste is mixed into pinole for an interesting take on the dish. Pumpkin seeds were once an ingredient for the Mexican drink horchata, which today is made with rice. Many Hopi and other Pueblo Indian cooks still season their piki stones with pumpkin seed oil before making piki bread. The nutrient-rich blossoms

Squash and pumpkins are, of course, cultivated and harvested primarily as food, but many native peoples save their seeds for other purposes.

(see previous entry) are eaten directly with soups and stews or dried for later use, mixed with cornmeal, mesquite, and meats. Young pumpkin and squash leaves make excellent and tasty greens when steamed, sautéed, or boiled.

Squash and pumpkin leaves are also taken to treat upset stomachs. The seeds of *Cucurbita pepo* are eaten by many native peoples to treat intestinal worms, as a diuretic, and as a kidney aid. The Zuni use the seeds of gray squash, a summer squash, as an ingredient in their *schumaakwe* cakes, a mixture applied externally for rheumatism and swelling, and they make a poultice of the blossoms and seeds to treat scratches from cactus. The Cheyenne take an infusion of pumpkin rinds as a laxative and kidney aid, as well as to treat rheumatism, arthritis, fevers, and earaches.

IDENTIFICATION AND HARVEST

Cucurbits are generally easy to recognize. Plants are vine-like with sprawling, prostrate stems emanating from a shrub-like central plant. Sometimes the vines climb using tendrils to catch onto nearby fences, ladders, and other plants. The large palmate leaves are arranged alternately on the stems; they are pinnately lobed and often have small thin spines on the petioles. The flowers are yellow and star-shaped (see previous entry); the fruits are hard-shelled and are produced in a variety of shapes and sizes. Some farmers have grown 900-pound squashes and 1,800-pound pumpkins. The soft-skinned summer squashes include such varieties as straightneck, crookneck, baby acorn, chayote, zucchini, patty

Alamos Cushaw, an heirloom squash.

pan, and scallopini. The hard-skinned winter squashes and pumpkins include acorn, butternut, delicata, golden nugget pumpkins, carnival, spaghetti, sugarloaf, and turban. Besides *Cucurbita pepo* (winter and summer squashes, pumpkins, ornamental gourds), the chief North American cucurbits are *C. mixta* (winter squashes, pumpkins, cushaws) and *C. moschata* (winter squashes, pumpkins). The different species are distinguished by their skin hardness and stem shapes.

Summer squash are typically ready for harvest from middle to late summer and into the early fall. Winter squash are grown to be harvested later in the fall, once the shells are hard enough to withstand a little pressure from a pressed fingernail. The seeds are often scooped from the insides of winter squash before cooking. They will be intertwined with stringy flesh and pulp. I often lay the seeds and stringy pulp out on paper towels or newsprint in an area protected from insects; it will all dry in a couple of days, and then the seeds are more readily separated from the pulp. When collecting squash or pumpkin leaves for consumption, be careful of the spines situated on the stems. It is also important to split open the thick hollow petiole and to then pull up toward the top of the leaf in order to remove the hard veins from the leaves before cooking.

HEALTH BENEFITS

Curcurbit flesh is an excellent source of nutrients, full of carotenoids, lutein, soluble and insoluble fiber, potassium, vitamins C and E, B complex vitamins (including riboflavin and thiamine), and magnesium. Curcurbit seeds contain many phytochemicals, including phytosterols, polysaccharides, peptides, and fatty acids, as well as squalene, which is effective in the treatment of certain types of cancer. Together, the constituents improve urinary function and are anticancer, antioxidant, antihyperglycemic, antiulcer, anti-inflammatory, and antimicrobial.

STAGHORN SUMAC
RHUS TYPHINA

Family: Anacardiaceae
Parts Used: whole plant
Season: summer, fall
Region: eastern North America

According to the USDA's Natural Resources Conservation Service, staghorn sumac can be weedy or invasive. How can a plant be deemed invasive if it is native to North America? Staghorn sumac thrives in a variety of habitats, including roadsides and other disturbed sites. It can form large colonies in the dry, rocky soil along abandoned fields. It sprouts easily and grows rapidly, which can eliminate or reduce the abundance of other species that cannot persist in the shade it creates. But from a native person's perspective, staghorn sumac is neither a weed nor an invasive plant. It is considered an important and useful plant for medicine and for food.

USES

Many eastern tribal nations use the little red fruits of sumac to create a tart drink. The fruits are collected, soaked in cold water, strained, and then sweetened and made into a pinkish drink whose flavor is reminiscent of lemonade. Staghorn sumac is related to Sicilian sumac (*Rhus coriaria*), a common spice in Mediterranean and Middle Eastern cuisine. Some native peoples will mix the leaves and berries of staghorn sumac with tobacco and other plants as part of their ceremonial smoking mixture. Algonquian-speaking people make an infusion of the plant to treat rheumatism and to improve the appetite. The Cherokee eat the berries to stop vomiting and use an infusion of the berries to treat sunburn. An infusion of the bark is taken to help mother's milk flow. The Chippewa take a decoction of the flowers for stomach pain, while the Delaware take an infusion of the berries to treat diarrhea. The Menominee boil down a decoction of the berries for coughs and drink a decoction of the berries for pinworms. The Ojibwe use an infusion of the root to treat hemorrhages.

All parts of staghorn sumac can be used as both a natural dye and as a mordant. The plant is rich in tannins and can be added to other dye baths to improve lightfastness. The inner bark makes an orange dye, the roots are boiled for a yellow dye, and the berries make either a black or red dye.

Staghorn sumac is considered an invasive species by many contemporary plantsmen, but it has always been useful to native peoples.

IDENTIFICATION AND HARVEST

Staghorn sumac is a tall, deciduous perennial shrub/tree, up to 30 feet tall, that usually grows in colonies. The twigs are often crooked and velvety. The pinnate-compound leaves are large (16 to 24 inches in length), bright green on top and pale underneath; the leaflets are lanceolate and serrate. The leaves will become extremely colorful in early fall. On female plants, each tall pyramidal inflorescence of fuzzy yellow-green flowers becomes a cluster of bright, fuzzy red berries that persist through winter. The berries are small, less than a quarter-inch in size. The size of staghorn sumac—in addition to the clusters of red fruit in season—make it an easy plant to find. Look for staghorn sumac throughout the eastern half of North America. It tolerates a variety of

Staghorn sumac berries are used to create everything from refreshing beverages to medicinal infusions and decoctions.

conditions but is most often found growing in open fields, roadsides, along fences and railroad rights-of-way, and in burned areas. It is not really tolerant of shade.

HEALTH BENEFITS

Eleven different types of flavonoids have been identified in staghorn sumac, which accounts for the antibacterial, antifungal, and anthelmintic effects of this plant. Its fruit contains phenolic acids and anthocyanins. Pyranoanthocyanins and other polyphenols are responsible for its antioxidant and anti-inflammatory activity.

STINGING NETTLE
URTICA DIOICA

Family: Urticaceae
Parts Used: whole plant
Season: year-round
Region: North America

For a while now, loaded with a 60-pound backpack, I had been moving through the forest. As I dropped down into a small ravine, I was engulfed in the shade created by the towering redwoods and coastal spruces, which contribute so much to the habitat of northern California's Lost Coast Wilderness. The ravine had been carved by a narrow stream, whose water flowed around small rounded stones and over mosses, watercress, and other aquatic plants. I carefully identified a series of larger dry rocks that I planned to use as stepping stones

Some native fishermen rub the leaves of stinging nettle over their lines to turn them green and to mask their human scent.

is taken to relieve fevers, to help stop the blood from flowing following childbirth, and to treat rheumatism. An infusion of the roots is also taken to treat rheumatism, as well as to relieve upset stomach, hives, and rashes. The leaves are taken as an astringent, as a diuretic, and as a general tonic. Many native peoples collect the leaves and cook them as an edible green. The tiny "stingers" disappear when cooked. The whole plant is rubbed on the skin to treat aches and pains, skin infections, rheumatism, and arthritis. The central stems of stinging nettles are pounded or twisted in order to make string cordage for bow strings, fishing nets, basketry, ropes, and twine. Some native fishermen rub the leaves onto their fishing lines in order to mask their human scent and to turn the lines green.

IDENTIFICATION AND HARVEST

Stinging nettles are herbaceous perennials that can grow up to 6 feet tall. The bright green leaves, up to 7 inches long and with stinging hairs, are narrow-lanceolate (sometimes wide and ovate) and arranged opposite each other on the central stem; the base of the leaves tapers to cordate. The individual plants are topped with a spike-like inflorescence containing several panicle-like pistillate flowers. The small fruits are ovate. The whitish tan roots are rhizomatous; they are about a half-inch wide at the top nearest the aboveground part of the plant and several inches long. Look for growing and mature nettles from March to November. The plant will be in flower from May to October. The plant attracts elk and other wildlife.

in order to cross the stream while keeping my boots dry. I successfully managed the first two rocks and kept heading toward my ultimate destination, a thick wall of green shrubs and vines on the opposite bank. As I approached, I was stopped midstream by an unusual sound emanating from the tangle of green before me. I moved another rock or two closer—and came face-to-face with a large snorting Roosevelt elk. It stared back at me, nonplussed by my presence, and casually continued chewing on a streamside stand of stinging nettles.

USES

Like so many plants favored by American Indians, one stinging nettle plant can have several purposes: as a medicine, as a food, and as a material for making utilitarian objects. An infusion of stinging nettle leaves

Stinging nettle roots are used to create a medicinal infusion useful for treating upset stomach, hives, rashes, and rheumatism.

It grows near streambanks that are in partial sunlight. Stinging nettle can also be found in disturbed areas along trails and roadsides throughout North America. Wear gloves and protective clothing when gathering the plants. The stinging sensation results from contact with the hairs on the leaves; any associated pain can be alleviated with heat.

HEALTH BENEFITS

Stinging nettle contains formic acid, galacturonic acid, ascorbic acid, choline, flavonoids, terpenes, acetylcholine, vitamins A and D, iron, potassium, sodium, phosphorus, calcium, silica, and albuminoids. The stinging hairs have high concentrations of histamine.

STRAWBERRY
FRAGARIA SPP.

Family: Rosaceae
Parts Used: leaves, fruit, roots
Season: spring, summer, fall
Region: North America

The roundhouse ceremonies and annual festivals of thanks and renewal had ceased for quite a while among native peoples of California. Before World War II, the communities were still rebuilding from two centuries of near-genocide. During the war, many of the remaining inhabitants had moved away to find work in the orchards or in the hop fields, or to help with the industrial war machine. After the war, the Kashia Pomo community, in an effort to instill harmony with outsiders, decided to ignore tradition and open their roundhouse ceremonies to nonnative visitors; it was at this time that they began to celebrate their annual strawberry dances.

For many American Indian communities, strawberries represent annual spring renewal. Some Algonquian-speaking tribes feel that strawberries are symbols of thanksgiving. Each June the annual Kanatsiohareke Mohawk Strawberry Festival is celebrated with a pow wow, food, music, crafts, and storytelling at the Kanatsiohareke Mohawk Community in Fonda, New York. Strawberries are even part of the pantheon of sacred Navajo life medicines. The Cherokee believe that strawberries are a love medicine, originally given to us by the Creator in order to reunite First Man and First Woman after they had been quarreling. The story tells how after a dispute, First Woman left her husband and headed east, toward the Sun. First Man was very sad and began to follow his wife. The Creator looked down upon First Man with pity and asked if he were still angry with First Woman. When First Man confessed he was no longer upset, the Creator told First Man that he would help him get his wife back. The Creator caused huckleberries to ripen along the path that First Woman was following, hoping to slow her progress as she stopped to collect them. But she was not interested in them. The Creator tried other kinds of sweet berries—blackberries, serviceberries—but nothing caused her to pause, much less stop. Finally the Creator created a new kind of bright red berry and placed it in her path. She finally stopped to eat some of these new berries. As she was eating, she glanced west. This caused her to think of her husband and to realize that she missed him. She gathered more berries to bring to him, turned around, and headed west, toward home. First Man and First Woman met each other on the trail and were happily reunited while sharing the newly created strawberries.

Wild strawberries are featured in legends as symbols of spring, renewal, thanksgiving, medicine, and love.

USES

Many American Indian tribes across North America consider strawberry (e.g., *Fragaria vesca*, *F. virginiana*) to be an important fruit. It is the first fruit to ripen in the spring. But eating those wonderfully tart and sweet berries is not their only use. The Navajo believe strawberry is good medicine because it is shaped like a heart; when it is eaten or when the juice is drunk, we are revitalized. Many native peoples treat diarrhea and dysentery with a decoction or infusion of the leaves and roots; the same is also used to relieve gout or an inflamed bowel. The leaves are dried and powdered and applied to open wounds as a disinfectant. The powdered leaves can also be made into a poultice and applied to sores on the skin. Strawberry juice is taken as a general tonic and to relieve arthritis.

IDENTIFICATION AND HARVEST

Wild strawberry is an herbaceous perennial that normally spreads through short rhizomes. The plant grows up to 4 inches tall. The thin green trifoliate leaves, about 3 inches long and half as wide, are serrate. Strawberry plants produce small white flowers with five petals and a yellow center. The fruit of the wild strawberry is much smaller than that of the hybridized or store-bought variety. Woodland strawberry (*Fragaria vesca*) most resembles the hybridized garden-variety strawberry that many people purchase from the local nursery, except on a much smaller scale. Look for strawberries in cool, moist, shaded areas near the edges or in open areas of forests. Strawberries grow throughout North America. When collecting the leaves, do not gather the entire plant. When gathering the roots, be sure to leave some of the root system in the ground for future growth.

HEALTH BENEFITS

Strawberry leaves and roots contain pedunculagin and other tannins, which play a role in its use for inflamed bowels; they also contain such anti-inflammatory and antibacterial agents as ascorbic acid, quercetin, and salicylic acid.

SWEETGRASS
HIEROCHLOE ODORATA

Family: Poaceae
Parts Used: whole plant
Season: summer, fall
Region: North America

On the campus where I teach courses in American Indian studies, dried sweetgrass braids are displayed on my office desk and bookshelves. The braids are gifts I have received from native colleagues and students. One of them is burnt at the end from the time when, risking setting off the building's fire alarm, I demonstrated to a class what this ancestral American Indian incense smells like. It is difficult to describe the light, sweet, and pleasant scent of sweetgrass, both when it is fresh and when it is dried and then burnt. The closest I can come is this: imagine the scent of vanilla bean wafting through the air, coupled with burnt marshmallow on a stick, all while sitting around a campfire.

Every tribe that relies on sweetgrass for incense maintains a story for how the plants first came to the people. Among the Blackfoot near the headwaters of the Missouri River, it is said that at one time a young Blackfoot man whose face bore a terrible scar from a past incident approached a beautiful girl, hoping to win her affection. When she laughed at him,

A traditional sweetgrass braid.

he ran away, far to the south. Weary from running and from lack of food, he fell asleep in the deep prairie grasses. Morning Star, the son of the Sun and Moon, felt sorry for the young man and, with his parents' permission, transported him up to the sky world. Using a forked stick to hold a hot ember, Morning Star burnt sweetgrass around the young man. This was done in four consecutive sweat lodge healing ceremonies. After the fourth time, the scar disappeared from his face. When the young man returned to his home back on the earth, it is said that he shone with a yellow light. He taught his people the sweetgrass purifying and healing ceremony that he had learned while in the sky world. To this day, the Blackfoot employ the sweetgrass ceremony in their sweat lodges as it was taught to them by the young man.

USES

The Blackfoot story of sweetgrass highlights the primary American Indian use of this plant: it is burnt as an incense for purification and for spiritual, emotional, and physical healing.

Some Northern Plains tribes will also inhale the smoke of sweetgrass to ease colds and drink an infusion of the plant for coughs. The stems of the grass are boiled or soaked in water and used as a wash for rashes and for chapped skin. Some ceremonial participants who have been fasting will chew on the leaves for endurance. The leaves are also chewed for sore throats. For those who are in search of natural mosquito repellents, know this: native peoples use it as such by keeping sweetgrass on their person and in their homes. The long

Sweetgrass is famous for its use in basket making.

Sweetgrass bundled into a wreath in preparation for braiding.

fresh stems of sweetgrass are easily braided; some Northeast Indian peoples will braid and weave sweetgrass with splints of ash into intricate baskets.

IDENTIFICATION AND HARVEST

Sweetgrass is a perennial grass of wetlands and riparian ecosytems. It grows in thick rhizomatous mats in moist meadows, along the edges of ponds and streams, and even near the ocean. The plant emits a sweet scent. The stems grow up to 3 feet tall and produce a tapering branched inflorescence, with clusters of flowers arranged in panicles. The stems are lightly purple near the roots. Sweetgrass can be found growing across the northern third of the continent, from Alaska to the Canadian Maritimes, and south to Virginia in eastern North America, but it also appears in scattered regions in the Southwest. Harvest time for sweetgrass begins in late July and extends into September.

The best time of day to harvest sweetgrass is during the early morning, just as the sun is stretching its warm glow onto the plants.

Sweetgrass in flower.

Sweetgrass often grows with other grasses and plants. Run your hand through an area of several grasses, and the sweetgrass will sort of glow a brighter green than the other grasses in the area.

Because sweetgrass is a sacred and ceremonial plant, it is important to honor it with a gift before harvesting. Native harvesters will offer a verbal thank-you, a song, or a sprinkle of tobacco or cornmeal prior to harvesting. It is also important to properly harvest this plant. Do not pull up the entire plant; this will damage the root system and deplete the population. It is better to snip the plant near the base, leaving the roots in the ground; this ensures that plants will return the following year.

HEALTH BENEFITS

Sweetgrass contains coumarin and phytol, both of which phytochemicals repel biting insects, as well as benzopyranone, which acts as an antioxidant.

THISTLE
CIRSIUM SPP.

Family: Asteraceae
Parts Used: whole plant
Season: spring, summer
Region: North America

One of my students asked with disbelief, "How do you eat a thistle?" It's understandable that thistles would be avoided by many. They are perceived as too difficult to deal with due to their prickly spines or are categorized as noxious or invasive plants. Across the continent American Indians, however, have long regarded thistles as useful plants for both food and medicine. On the tallgrass prairie and in other locations where native grasses, plants, and wildlife are undisturbed, they are keystone species.

USES

The Lakota call thistle *thokahu*. They peel the root and young stems and eat them raw or cook them in soups and stews. A decoction of the root is used to treat gonorrhea, and an infusion of the root is dropped into the eyes to clear infections. Other native communities eat the roots of thistle as well. They can be roasted over coals, boiled, or pit-roasted and stored for later use.

The Seminole make the stems of *Cirsium horridulum* into blowgun darts. In the Southwest, *C. neomexicanum* is used as an eye

Thistle's prickly reputation has kept many from using it for food and medicine.

medicine. An infusion of the root is dropped into the eyes. An infusion of the entire plant is taken for chills and fevers. The Kiowa use a decoction of the blossoms of *C. ochrocentrum* as a burn dressing. The root is also used as a contraceptive, and the whole plant taken as an emetic to treat syphilis. The Chippewa take a decoction of the root of thistle to treat back pain. The Cherokee use an infusion of the leaves of *C. altissimum* as a general analgesic for nerve pain. A poultice of the roots is used to treat sore jaws. A decoction of thistle roots can also be used to remove worms in children. The roots can be used for stomachaches and the whole plant as a laxative. The dried powdered stem is sprinkled onto wounds to dry and disinfect them.

IDENTIFICATION AND HARVEST

Thistles may be annuals, biennials, or perennials. They are known for their prolifically produced composite (ray and disc) flower heads. The radially symmetrical flowers are terminal, borne at the ends of branches originating from a centralized woody stem, which arises from a large rosette of prickly basal leaves. Plants also have deeply lobed prickly stem leaves. Species vary, but in general leaves are alternate and sometimes slightly hairy, and flowers are purple, pink, red, yellow-orange, or white.

Tall thistle (*Cirsium altissimum*) grows up to 13 feet tall. Its leaves are more toothed than spiny and have shallow lobes. The flower head is pink to purple. Clustered thistle (*C. brevistylum*) is an annual or biennial growing up to 6 feet tall. It normally grows as a single-stemmed

Cirsium occidentale var. *venustum*, about to bloom; note the cobwebby fibers, which help to separate it from many other thistles.

plant. The leaves are deeply lobed and contain twisted spines. The top of the plant and flowers seem to be covered with webby fibers. The flowers are white to pink. Like those of clustered thistle, the flowers of cobweb thistle (*C. occidentale*) are heavily covered with fine fibers that resemble cobwebs. This thistle usually grows in low clumps, but plants can occasionally reach 9 feet in height. The leaves are gray-green, toothed, with triangular lobes. The flowers can be a stunning blood-red, maturing to purple.

North American thistles grow in many kinds of habitats, from the open parts of forests and woodlands to prairies, roadsides, disturbed areas, and (in the case of clustered thistle) coastal scrublands. They prefer areas where the soil is moist. If the stems are intended for eating, they should be harvested in the spring to early summer, before they become too fibrous and woody. Take care when collecting the leaves and flowers, as the spines and teeth can painfully break the skin. Leather gloves and pruning shears are recommended.

Cirsium species contain beta carotene, fatty acids, vitamin C, and nutritional minerals. The genus also contains scopoletin, pectolinarigenin, and acacetin, which demonstrate antibacterial and antifungal activity. Native thistles contain coumarin, flavonoids, phenolic acids, tannins, sterols, triterpenes, phenolic glycosides, and monoterpene lactones; these compounds are known to be diuretic, astringent, and anti-inflammatory but can also damage the liver over a period of use.

TOBACCO
NICOTIANA SPP.

Family: Solanaceae
Parts Used: leaves
Season: spring, summer, fall
Region: North America

For as long as we can remember, tobacco has been a sacred plant to native peoples. No other plant holds as much universal importance among the indigenous peoples of North and South America. It is understood that it must be used with respect and plays a central role in maintaining our ways of life, our identities, and our relationships to our landscapes. Each tribe across the Americas maintains its own narrative regarding the origin of tobacco, how it came to the people, and how to use and care for the plant; but it is generally believed that, when we use tobacco in ceremony and ritual, the smoke captures and delivers our thoughts and prayers to the natural world around us and on into the spirit realms. This concept is directly related to the title of this book: one of the meanings of iwígara is "breath," and the general American Indian concept of breath is similar to that of soul, or spirit. Breath permeates all things and the universe. And humans share that breath. When we inhale the smoke of tobacco leaves, we are creating a unifying connection with all the cosmos. The smoke flows in and from our bodies, carrying with it our thoughts, further solidifying our connection to all things.

USES

Tobacco is often offered as a gift before collecting plants or hung in bushes and trees as a gift to the spirits of a sacred location. Tobacco

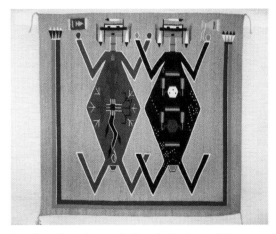

Navajo sandpainting rug by Brenda Crosby depicting Mother Earth and Father Sky. Note the four plants that become the cardinal directions woven into the body at left; tobacco is depicted as the plant to the east.

is used by my people, the Rarámuri, and by the Paiute as a snakebite remedy. The leaves are chewed into a poultice and applied to the area of the bite. The Rarámuri, as well as the Navajo, apply a poultice of the leaves to the forehead to relieve headaches and to other areas of the body where there is pain. The poultice is wrapped in moist cloths and applied to the area of pain. My people smoke *Nicotiana rustica*, *N. glauca* (introduced to North America from Peru after European contact), and *N.*

obtusifolia (the latter only at night). The smoke is also used for the blessing of ceremonial objects. *Nicotiana glauca* is now naturalized in the Southwest, where it is used by many other native communities, not just the Rarámuri. *Nicotiana attenuata* is used by many tribes in the Southwest to stop nosebleeds and as a wash to heal wounds, pains, and bruises. A very watered-down decoction of the leaves is sometimes used as an emetic and to rid the body of worms. Hispano herbalists in New Mexico will

Nicotiana glauca features striking yellow flowers—and a nicotine concentration much higher than that of commercial varieties.

An 18th-century botanical illustration of what is now *Nicotiana rustica*, showing the iconic leaf form of this genus.

Nicotiana quadrivalvis features pure white flowers.

mix the powered leaves with lard and place the poultice, layered between linen cloth, onto the chest of patients suffering with a chest cold. California Indians use *N. clevelandii* as a poultice to treat cuts, bruises, and wounds.

Tobacco is always used as an external medicine. It is extremely dangerous if consumed internally. During one of my university ethnobotanical field schools, my students and

I had driven our two vans into the canyon area fed by the Salt River east of Phoenix. We stopped at a pullout so that I could show the students a large specimen of *Nicotiana glauca*. This particular plant must have been 15 feet tall. I described the various uses of the plant, making the point that caution should always be taken with wild tobaccos, as their nicotine content is several times higher than that of the domesticated *N. tabacum*, which is hybridized by cigarette companies. We returned to the vans and continued our journey east. About an hour after the stop by the tobacco plant, an urgent call came over the walkie-talkies that we were using to communicate between vans. The students in the other van were telling me to stop: one of their fellow students was sick. When I reached the ailing student, she was vomiting, her skin was reddish and feverish, and her lips were turning black. She had taken one of the leaves from the *N. glauca* plant and begun to chew on it. We were at least an hour's drive from any kind of hospital, and no one's cell phone had a signal. I forced lots of water into the student, trying to induce more vomiting. After about 20 minutes she began to return to normal, but she was not herself for at least another day.

IDENTIFICATION AND HARVEST

Tobacco is an annual; plants are usually about 2 feet tall, although some species in good conditions can reach up to 5 feet in height. In the case of *Nicotiana glauca*, plants can be as much as 15 feet tall. The leaves are alternate and carried on long petioles; they are entire, ovate to lanceolate, and are often longer near the base of the plant and reduced gradually in length toward the top. Both the stem and leaves are pubescent. Most species sprout pale yellow, trumpet-shaped flowers, about an inch long, on terminal racemes or panicles. The flowers of *N. longiflora* and *N. quadrivalvis* are white; those of *N. tabacum* are pink. All tobacco flowers are five-lobed.

Certain *Nicotiana* species are native throughout North America. They often emerge in areas that were burned during the previous year. The plants seem to appear in places where the collector would not expect them to be. In the West, tobacco does not grow above 7,000 feet in elevation. Tobacco does well in dry areas but to do so will require a moist spring season. Look for stands of plants in flats, meadows, and dry streams. The leaves are the useful part; do not pull up entire plants, but rather snip individual leaves to be used fresh or dried for later use.

HEALTH BENEFITS

Tobacco contains many pharmacologically active compounds, including the alkaloids nicotine and nornicotine. Alkaloids are antitumor, narcotic, analgesic, and antimicrobial. Other active compounds include guaiacol, quercetin, rutin, and eugenol. Guaiacol is an antituberculic and expectorant. Quercetin is an antispasmodic, diuretic, vasodepressor, and viricide. Rutin is antiatherogenic, antiedemic, anti-inflammatory, antithrombogenic, and hypotensive. Eugenol is fungicidal, anesthetic, antiseptic, and larvicidal.

TULE
SCHOENOPLECTUS ACUTUS
BULRUSH

Family: Cyperaceae
Parts Used: whole plant
Season: year-round
Region: North America

In the West, and especially in California, tule is more than a tall kind of water rush found in many lakes and ponds around the region. For many people, especially California Indians, tule is connected to cultural heritage, worldview, and even identity. There are tule elk, tule ponds, and even something called tule fog. Northern California has the Tule Lake National Monument (which includes the internment camp), a Tule Lake Elementary, and Tule

The seedheads of tule.

A Pomo shelter constructed of tule, circa 1924.

Lake High School. The Tule River Tribe, with around 1,800 currently enrolled members, are among the original indigenous inhabitants of California's San Joaquin Valley; they are Yokuts-speaking people who traditionally occupied areas along the rivers, creeks, and other tributaries of Tulare Lake between the Sierra Nevada and the Coast Ranges.

USES

Native peoples from California and the Basin and Range region of the West maintain a long-standing relationship with tule. Tule can be eaten, is used for various utilitarian purposes, and has some medical value. The young shoots of tule that emerge in early spring are eaten raw or can be cooked. Pollen from tule flowers is collected in baskets or pails and used like a flour for breads, cakes, and mush. Later, as the plant matures and the flowering spikelet "fruits," the small seeds are beaten into a bucket or basket and ground into a meal. The large roots of tule can be eaten raw, cooked, or dried for later use, when they are then pounded into a meal. Medicinally, the stems of tule are used by the Cree as a dressing to stop bleeding. The Navajo take the plant as an emetic.

Several California Indian tribes who were historically situated near the Bay Area and large lakes made canoes of tule stems; the long stems were bound together with vines of wild grape. Some native groups situated near mud flats and around the coast made large round shoes from tule; the shoes would permit them to traverse muddy areas and marshes. Some West Coast peoples made clothing from tule; during the rainy season shredded tule hats and cloaks repelled the rain. Native peoples around Tulare, Mono, and other lakes still construct duck decoys from tule. Thanks to its marked insulating properties, tule is often woven into sleeping mats and roofing materials. Native California and Nevada Indians, such as the Yokuts, Pomo, Ohlone, Miwok, and Paiute, even construct small dwellings out of bunched-together tule. The shelters provide insulation and are rain-proof. The structures are framed and anchored to the ground with willow poles and lashed together with tule or cattail cordage; the walls are thatched with mats of tule or cattail and secured to the frame.

Tule is used by native basket weavers throughout the West. The stems and leaves are woven into baskets with lids, and even waterproof containers. The long rhizomes of tule are collected and used for the dark brown and black design elements of some baskets. The rhizomes are dried and then soaked for several months with acorns, walnut shells, ash, or other such materials, in order to create the dark brown or black color.

IDENTIFICATION AND HARVEST

Tule is a deciduous herbaceous plant with cylindrical central stems up to 8 feet tall, sheathed by long slender V-shaped leaves. An orange-brown flowering spikelet forms at the top of the stem; the inflorescence includes at least three panicles of spotted, tightly formed flower bracts containing scaled seeds. Tules grow in large clonal groups. The roots and underground rhizomes are long, thick, and stout. Look for tule in wetlands, lakes, ponds, and freshwater marshes throughout the West. It is often intermixed with cattails. The stems can be harvested at the base in order to protect the root and rhizome system. Be prepared to get wet if you intend to harvest the roots.

HEALTH BENEFITS

The nutritional value of tule is similar to that of cattail. Plants contain several polyphenols, cinnamic acids, and flavonoids; the tannin catechin may play a role in the hemostatic activity of tule stems.

VERBENA
VERBENA SPP.

Family: Verbenaceae
Parts Used: leaves
Season: spring, summer, fall
Region: North America

Most plant species have co-evolved with particular ecosystems and habitats. Some plants, such as the columnar saguaro or majestic redwoods, are endemic to specific regions. Similarly, you will not see mountain mahogany growing east of Oklahoma, nor will saw palmetto be found growing wild in the Mountain West. That's why I'm constantly reminding

Verbena has long been used as a calming agent. Pictured here is *Verbena stricta*.

The flower spikes of *Verbena hastata*.

my ethnobotany students that all American Indian knowledge is local. Native knowledge of plants is tied to the ecosystems of their traditional homelands. As a result, only a handful of plants were ethnobotanically known to nearly all pre-contact American Indians; the native verbenas are a part of that small group.

USES

Dormilón ("they slept," "sleepyheads"), the Spanish name for verbena, describes one of the uses for this plant, all of which involve the leaves. An infusion of the leaves is given as a sedative for insomnia and stress and to children who are fidgety prior to the onset of a cold or the flu. The infusion is also taken to relieve upset stomachs and nasal congestion, to settle coughs that result from upper respiratory infections and inflammation, and as a diuretic. An infusion of the leaves can also be consumed to help muscle and skin tissue better heal from sprains and contusions. The leaves can be placed inside the mouth to relieve toothache.

IDENTIFICATION AND HARVEST

Verbena is an herbaceous square-stemmed annual that can grow up to 4 feet tall. The upper part of the plant's main stem normally ends in several spikes. The upper half of the spikes are covered with half-inch purple, lavender, or bluish flowers that are arranged in rings around the spike; some species (e.g., *Verbena macdougalii*) can have whitish flowers. Plants flower in July and August. The flowers are usually five-petaled and are fused at the bottom to form a short tube. The 2- to 3-inch opposite leaves are serrate and ovate, with pointed tips. Both leaves and stem are covered with dense hairs. Verbena grows wild in most of the contiguous United States and Canada. It is usually found in dry sandy soils, in full sun, and often invades overgrazed pastures, meadows, and disturbed fields; it does not do well around native grasses. Harvest leaves while the plant is in flower. Avoid collecting the woody stem.

HEALTH BENEFITS

Verbena contains the secoiridoid glycoside verbenoside and the iridoid glucoside verbenalin. Tannins and flavonoids (luteolin, kaempferol, apigenin) are also present. These compounds are also antibacterial. Verbenalin is anti-inflammatory, analgesic, sedative, anticonvulsant, and anxiolytic.

WAPATO
SAGITTARIA SPP.

Family: Alismataceae
Parts Used: leaves, roots
Season: spring, summer, fall
Region: North America

I often wonder what went through the minds of Meriwether Lewis and William Clark back in October of 1805. They and their Corps of Discovery, composed of 32 military recruits, had just barely survived the westward crossing of the Continental Divide by following their guide, Sacagawea, who had given birth earlier that year, in February. They were unfamiliar with the territory of the Bitterroot Mountains and the Lolo Trail and, therefore, not equipped to forage the land for edibles. They survived the crossing by eating candles, their horses, and something called "portable soup." They were rescued on the Clearwater River and restored to health by the Nez Perce, who fed them wapato, one of their staple foods. Wapato is a tuber that reminded Lewis and Clark of small potatoes. They and their men ate heartily of this new food. But the next day Lewis and Clark and the entire Corps of Discovery were laid up with dysentery. Perhaps the Corps of Discovery had fallen victim to the glycosides present in wapato? Nez Perce and other tribal peoples of the area had co-evolved with the tubers for many generations. They no longer had a negative reaction to the slightly toxic glycosides; however, Lewis and Clark and their men were not only not used to the toxic compounds, they

Wapato tubers, a nutritious food source for the Nez Perce.

collected to be dried and then reconstituted for consumption. The dried tubers are pounded into a flour and then added to soups and stews or baked into cakes. Wapato is also a medicine. An infusion of the leaves is rubbed onto a person suffering from a fever and taken to treat rheumatism and headaches. An infusion of the roots acts as a laxative and is taken for indigestion. The Potawatomi apply a poultice of the tubers to wounds and sores. Finally, wapato once acted as a trading currency among Pacific Northwest peoples, and the tubers are used for gambling games.

also ate heartily as a result of being nearly starved. In any case, the Corps could not continue their journey west for several days. Again, I wonder—what were they thinking?

USES

The small tubers of wapato are eaten by indigenous peoples across North America. *Sagittaria latifolia* is found in most of the continent; *S. cuneata* occurs primarily in the Pacific Northwest. The tubers are cleaned and boiled or roasted before being eaten. They are also

IDENTIFICATION AND HARVEST

Wapato is an aquatic perennial, with arrowhead-shaped alternate leaves arising from a rhizome. Each leaf grows on one stem. The leaves of *Sagittaria latifolia* are up to 12 inches long; those of *S. cuneata* are 2 to 6 inches long. The leaves are erect, though sometimes they rest on the water. Plants can grow

The three-petaled white flowers of *Sagittaria latifolia*.

The arrowhead-shaped leaf of *Sagittaria cuneata*.

up to 3 feet tall. Flowers bloom from July to September in sets of three on a central stem. Each flower has three white petals, each nearly an inch long, with many yellow stamens. The walnut-sized tubers grow on rhizomes about 3 feet below the bottom of the pond or stream of the plant's aquatic habit. The tubers have a white to bluish color on the outer layers and contain a milky juice inside.

Wapato grows in shallow pond edges, swamps, marshes, and slow-flowing creeks throughout most of North America and into Mexico. Leaves can be gathered whenever present, but the best time to harvest wapato tubers is in early spring or during the fall. To gather wapato tubers, reach below the surface of the water down into the mud or sandy bottom of the pond or stream. Follow the rhizomes with your hands, beginning from the center of the plant as the rhizome extends out. You will find the tubers growing at the end of each rhizome. Afterward, carefully clean the tubers before roasting, boiling, or drying. It is best to boil the tubers in salted water for about fifteen minutes. Before eating, remove the skin, as it is often bitter. The tubers could also be sliced into disks and strung on a line for drying.

HEALTH BENEFITS

Wapato tubers contain important nutrients such as protein, carbohydrates, insoluble fiber, thiamine, riboflavin, niacin, vitamin C, calcium, phosphorus, potassium, magnesium, zinc, and iron. They also contain diterpene glycosides, which are antifungal; the glycosides could be the cause of the Corps of Discovery's stomach issues.

WHITE CEDAR
THUJA OCCIDENTALIS

Family: Cupressaceae
Parts Used: leaves, branches, bark, wood
Season: year-round
Region: North America

White cedar is an important tree for many American Indians, valued for spiritual, practical, and medical reasons. It is native to the northeastern regions of our continent but can be found south to Georgia, and as far west as Illinois, Minnesota, and the Pacific Northwest. Whenever you are in the great forests of the Northeast and find a meadow, take a look at its edges, where the forest meets the open space. If you see a medium-sized conifer that looks like one of those classic Christmas trees that people manage to fit into shopping malls and into the tall foyers of office buildings, chances are you are looking at a white cedar.

The Potawatomi of Maine believe white cedar was, at one time, a human being. Apparently, a group of young men wanted to visit the Sun. They held a feast prior to leaving the village. They dressed themselves in their best buckskins and then set out, heading east. They found the place where the Sun rose, but as they approached, it got higher and higher in the sky. The men managed to get to the top of a mountain, where the Sun was not too far above. The next morning, they caught the Sun as it passed the mountain. The Sun was curious—why did the men go through so much trouble to reach it? The men told the Sun that all they wanted was to help their people with whatever power it was willing to grant them. They each requested and were given different powers that could be used to help their people when they returned home. One of the young men ask the Sun for everlasting life. In this way he would always be present in order to help his people. The Sun told the young man that upon his return home his request would be granted. On their way home, the young men were talking about their visit with the Sun. As they walked a stretch in single file, the man who had asked for everlasting life, who was bringing up the rear, stopped and announced that where they were at that moment was the place he was meant to stay. His companions looked back at him, but instead of seeing their friend, they saw a tall, wonderfully scented cedar tree. The tree spoke, telling the young men to take its leaves for ceremonies, for healing, and to use its wood for tools and other things, as it would be long lasting and not rot.

USES

Although native canoe builders have used different kinds of woods for gunnels and ribs of birch bark canoes, it is generally accepted that the best wood for this purpose is white cedar. The wood of white cedar is flexible, strong, and slow to decay or become waterlogged. With such qualities, it is no wonder that white cedar is also used for fence posts, cabin logs, poles, pails, barrels, tubs, basketry, and shingles. The Bella Coola of the Pacific Northwest include it in their first salmon ceremony. At the beginning of the salmon fishing season, when the first salmon are caught, the best example of the catch will be honored. Both red (*Thuja plicata*) and white cedar bark is used to construct a pillow for the fish as it's laid before a shrine and celebrated.

Cedar bark, wood, and boughs are burned as an incense during ceremony and blessings.

White cedar's iconic pyramidal shape stands out against other trees on the forest's edge.

The ribs and gunnels of this traditional birch bark canoe are made of white cedar.

The scent is believed to repel negative energy and to spiritually cleanse a space. White cedar is used for a variety of medicinal purposes. The Abenaki powder the leaves and apply it as a poultice for rheumatic swelling and pain. The Chippewa take an infusion of the leaves or inner bark as a cough medicine. Many tribes will inhale the steam of white cedar in sweat lodge ceremonies to treat colds, fevers, and rheumatism. A decoction of the leaves is consumed to treat urinary tract infections and applied to cuts, bruises, and sores. The boughs of white cedar are also used to repel moths and other pests.

IDENTIFICATION AND HARVEST

White cedar is an evergreen conifer, 40 to 80 feet tall, with a Christmas tree–like shape, pyramidal to conical and topped with a sharp crown. The dense and crowded branches spread toward the bottom of the tree, and the branchlets are flattened and fan-shaped. The fibrous bark of white cedar is gray to reddish brown and 0.5 to 1 inch thick; it is often composed of separate, flat ridges. The flat scaly leaves are pointed and are arranged in alternating pairs in four rows. They are bright green above, pale green below. Sometimes the leaves turn yellow-brown in the winter. The entire tree is often fragrant, especially when the leaves are crushed. White cedar produces hanging bunches of greenish brown seed cones that are generally ellipsoid and contain lateral wings. The seed cones are about a half-inch in diameter. Look for white cedar in partly shaded areas at the edges of meadows, near swamps, in rocky soils, and along the cool banks of streams. It can grow in dense groups and in sunny areas where there is plenty of moist soil. The young shoots of the leaves are best harvested as they are produced, late spring to early summer. The leaves and bark are available year-round.

White cedar produces greenish brown seed cones.

The leaves and bark are replete with essential oils, which are the source of white cedar's several immuno-stimulating and antiviral qualities. The main oil is thujone, a toxic agent that can harm the liver and kidneys and cause vomiting, stomachache, diarrhea, gastroenteritis, and chronic convulsions. But taken properly, thujone has beneficial pharmacological effects, being anthelmintic, anti-inflammatory, antirheumatic, antiseptic, antiviral, astringent, diaphoretic, and diuretic. It is also an emmenagogue.

Native American women knocking wild rice grains into their canoes with paddles, a traditional harvesting method in the 19th century.

WILD RICE
ZIZANIA SPP.

Family: Poaceae
Parts Used: fruit
Season: summer, fall
Region: North America

Many American Indians understand that we owe our very existence to the natural world around us. So many of our origin stories remind us of how certain animals, plants, and even stars in the night sky played a direct role in our emergence. In some cases, a specific food might not have played a role in the emergence of a people but has been essential to their ongoing existence and identity. A part of Ojibwe history includes a time when the people were forced to migrate west from their original homeland on the northeastern coast of Turtle Island. They were informed by a prophet that

they should travel and not stop until they came to land where food grew on water. The people followed these instructions until they reached the area around the Great Lakes. One day a hunter was returning to his camp downcast, without any game. When he reached his camp, he found a duck near his cooking pot. The duck flew away as the hunter approached but left some unknown grains in the pot. The hunter did not remove the grains left by the duck and made some soup anyway. He discovered that it was the best-tasting soup he had ever had. The next day the hunter traveled in the direction that the duck had flown. He reached a lake and found ducks and many other water birds feasting on the grains of some tall grasses that grew on the water. The hunter brought the new food to his people. After that the people understood that they would never go hungry so long as the gift of wild rice (*manoomin*), the lands, and the waters were taken care of.

Simply put, wild rice is wonderful food, cooked and eaten like any other kind of rice. But unlike white or brown rice, wild rice is not starchy. The inner grain is chewy. The overall flavor is a sort of sweet nuttiness. It cooks in about 30 minutes and does not absorb as much water as does white or brown rice. It is served as a main dish, or it can be part of soups, stews, stuffings, casseroles, and wildcrafted salads.

IDENTIFICATION AND HARVEST

Wild rice is not a true rice (*Oryza* spp.); that genus originated in Asia. Wild rice is an annual water grass that grows 3 to 6 feet tall. The elongated green ribbon-like leaves are 1 to 1.5 inches wide, with rough edges. The plant will sprout a broom-like flower cluster up to 2 feet in length and, eventually, at maturity, a single fruiting plume that contains several slender rod-like seeds. The seeds are tan to dark brown and

Zizania aquatica growing in South Carolina.

about a half-inch long; they are surrounded by a hull that terminates with a short beard.

In lakes and slower-moving streams in the Great Lakes region and northern Minnesota, the grains of wild rice begin to sprout underwater around May. The leaves of the plant will break the surface of the water beginning in June or July. The seedheads appear in late July and will begin to be ready to drop in August. In August and September, American Indians paddle their canoes into the wild rice areas to commence the harvest. The rare *Zizania texana* is native to the upper San Marcos River region of Texas; *Z. palustris* is native to the upper Great Lakes region, Wisconsin, Michigan, and up into Ontario.

HEALTH BENEFITS

Wild rice is a high-protein grain; a serving has a protein concentration of 12 to 15 percent. It's also high in soluble and insoluble fiber and an excellent source of iron, zinc, potassium, and other minerals; B vitamins like thiamine,

Wild rice is the basis of many wholesome and hearty dishes.

riboflavin, and niacin; antioxidant carotenoids; and cholesterol-lowering phytosterols, phenolics, and ferric acid. Its lipid content is lower than that of other common grains.

WILD ROSE
ROSA SPP.

Family: Rosaceae
Parts Used: stems, flowers, fruit, bark, roots
Season: spring, summer, fall
Region: North America

Roses are special plants for many native peoples. They symbolize life, persistence, and survival. Rose motifs are seen in bead and quill work and other forms of visual expression. They are mentioned in song. Some tribes believe that roses help to ward off ghosts. I always get a thrill whenever I spot wild roses during one of my backpacking trips or single-day forays into the woods. It seems that among all the wild relatives of cultivated plants, garden roses represent the extreme of what centuries of hybridization can yield. To come across a wild rose, then, is a special thing. I enjoy seeing their pale blooms facing the sunlight and appreciate the spots they have tenaciously carved out for themselves, often in less-than-ideal growing conditions.

According to an Anishinaabe legend, wild roses used to grow everywhere, in abundance. Their colors were vibrant, matching the colors of the rainbow. They were a favorite food of both animals and insects. Yet, they were taken for granted, neglected, and over a period of time, their numbers diminished. It was Bear, Bee, and Hummingbird who first took notice of the decline, especially when bears began to get too skinny, bees sick, and hummingbirds lost energy. Soon, it seemed there were no longer any roses around. A meeting of all the animals and insects was convened. Everyone offered their theories as to what had happened to the roses, but the creatures could not agree on a solution. Finally, it was decided that they needed to discover whether all the roses were truly gone. A search was begun, and all the swiftest birds were sent around the world to search for a rose. Finally, after much time, a hummingbird happened upon a lone rose barely clinging to a steep, rocky mountainside. The hummingbird brought the rose back and reconvened the gathering of the animals and insects. They asked the rose to explain what had happened to its relatives. In a weak voice, it said that the rabbits had eaten them

Encountering a wild rose always feels like a treat.

all. Immediately, the bears, wolves, and other large animals turned on the rabbits, grabbed them by their ears, and began to thrash them about the meeting space. This is why rabbits have long ears—they stretched out. Before all the rabbits were killed, however, the rose calmed everyone down. It told the gathering that it was not really the fault of the rabbits. The rose explained that if roses had been better cared for and not taken for granted, then their near-destruction would have been avoided. Afterward, roses began to inhabit the world again, but in smaller numbers and with paler colors. They also began to grow thorns in order to protect themselves.

USES

The several species of wild rose are used in much the same manner across our continent. The Lakota, for example, make a refreshing tea from the petals, hips, and roots of prairie rose (*Rosa arkansana*). Many tribes collect the hips of this and other species of wild rose and dry them for later use. The Blackfoot include rose hips in their pemmican recipe. The hips can also be roasted and eaten; they are often added to soups and stews. An infusion of the hips is taken for bladder infections and kidney stones. The same infusion can be cooled down and dropped into the eyes to relieve pink eye and other kinds of eye problems. The roots can also be crushed and made into an infusion for an eye wash. An infusion of the roots is also taken for sore throats, swellings at the joints, diarrhea, stomach problems, and dysentery. The bark is made into a decoction and taken

as an analgesic; an infusion of bark is taken for diarrhea and other stomach problems. The Cheyenne make arrow shafts from the stems and branches. Rose roots are pounded with gooseberry roots and juniper roots and woven into cordage.

IDENTIFICATION AND HARVEST

Wild roses are found throughout North America. Some species (e.g., *Rosa stellata*, *R. californica*) are endemic to smaller regions and pockets of the country. Wild roses are normally small deciduous shrubs, to about 6 feet in height. They can grow in dense thickets; the thickets are formed by the rhizomatous roots of the plants. Each plant produces many branches that sprout alternate, pinnately compound leaves; each finely toothed, obovate or elliptic leaflet is about 1.5 inches long. Flowers typically form in clusters on the newer growth in spring and summer. Each flower has small half-inch-long petals of pink, lilac-pink, or lavender. After the flowers, fleshy red elliptical fruits, or hips, are formed. The hips are collected in fall, after the flowers have mostly disappeared. Be careful while harvesting: dense thorns form on some species.

HEALTH BENEFITS

Wild roses contain useful amounts of phenolics, tannins, and flavonoids, and rose hips contain high amounts of ascorbic acid. These constituents demonstrate anti-inflammatory, antibacterial, analgesic, antioxidant, antimutagenic, and anticarcinogenic effects.

WILLOW

SALIX SPP.

Family: Salicaceae
Parts Used: leaves, branches, flowers,
    bark, wood, roots
Season: spring, summer, fall
Region: North America

It is not a traditional event for my people, but I enjoy going to the occasional pow wow hosted by other tribes. I relish the drumbeats and various singing styles. I welcome the mixing of the pan-Indian community and marvel at the athleticism of the young dancers. Every now and then there will be a special performance by hoop dancers. Hoop dancers move to fast drumbeats while manipulating up to six hoops at a time around, above, below, behind, and seemingly through their bodies. The regalia of hoop dancers is relatively simple and sparse, to permit the quick and agile movements in and around the colorful hoops, which are normally only about 2 feet in diameter and traditionally have been made of willow. In general, the dance and hoops represent the neverending cycle of life, a concept deeply respected and honored by American Indians. The hoop dance is also considered a healing ceremony by some tribes.

USES

American Indians put willow to many practical utilitarian uses. The flexible and resilient branches of the wood have long been a primary warp material in twined baskets and a foundation for coiled baskets. The materials can be manipulated in such a way that good weavers can construct hats, cooking vessels, serving bowls, trays, seed beaters, storage baskets, and baskets that hold water. Willow can even be made into twine that is thin enough for sewing. Willow wood can be woven into back rests and is used by native peoples for making weaving looms, arrows, blowgun darts, weaving sticks, and frames for ceremonial sweat lodges. Willow is considered an ideal soft wood to be

A Navajo performing a hoop dance, with hoops made of willow.

The flexible branches of *Salix alba*.

Weeping willow catkins, bright with pollen.

employed as a spindle stick for starting a fire with friction. Willow is used by the Hopi, Zuni, and other Pueblo Indians for prayer sticks. Wherever a strong, flexible wood is required, willow is ideal. Willow can also be eaten. The leaves of many willow species are used as a sort of spice in soups and stews. Young catkins too are added to soups and stews. The bark of new shoots is stripped off, and the tender shoot itself is then eaten. The inner bark can be cut into long slender strips, cooked, and eaten, or dried and pounded into flour to be used in cakes and breads.

Medicinally, an infusion of willow leaves and bark is an excellent analgesic, used by native peoples across North America to treat headaches, laryngitis, arthritis, muscle pains, and for many ailments where a pain reliever is required. The same infusion is used to reduce fevers and treat diarrhea and bladder irritation; it is used as a wash or made into a poultice to treat eczema and disinfect wounds, gangrene, and skin ulcers. Willow is also used as an anti-inflammatory to treat inflamed joints. A poultice of the bark and sap is applied to bleeding wounds. Willow roots are made into an infusion to treat aches and pains, diarrhea, and stomach problems. A decoction of the inner bark is taken to reduce coughs and relieve sore throats and colds. The Paiute will make a charcoal of willow bark and branches, add the ash to water, and take it to stop diarrhea.

IDENTIFICATION AND HARVEST

More than 100 different kinds of willows grow throughout North America. Some (e.g., black willow, *Salix nigra*) are large-trunked trees that can reach 85 feet in height. Others are medium-sized—for example, the slender *S. exigua*, which is about 25 feet tall. Still others are small willows (e.g., prairie willow, *S. humilis*), growing only to about 10 feet. Most are deciduous or semi-evergreen. The branches of willows are often slender, soft, and flexible; the bark is often gray or light in color and will feel damp. Willows are characterized by lateral, fuzzy, scaled leaf buds that emerge during late winter to early spring. The leaves are simple, feather-veined, and typically linear-lanceolate; usually they are serrate, pointed at the tips, rounded at the base. The leaves are variously colored, from light green and olive to yellow or even slightly blue. Willow roots are deep and large. Most often, willows are riparian plants found along streams, rivers, marshes, and general wetland areas. Prairie willow is the exception: it grows in upland areas in sandy soils and sparse woods in the Midwest. The best time to gather willow bark and wood is during the late fall and before the buds begin to form in the spring.

HEALTH BENEFITS

Willow's main pharmacologically active constituent is salicylic acid, one of the first phytochemicals to be isolated from a wild plant and manufactured synthetically into what we know as aspirin. Salicylic acid demonstrates analgesic, anti-inflammatory, antibacterial, antipyretic, antimicrobial, and antifungal activity. Plants also contain flavonoids, such as quercetin, which demonstrate pharmacological activity similar to that of salicylic acid.

YERBA MANSA
ANEMOPSIS CALIFORNICA

YERBA DEL MANZO, LIZARD TAIL

Family: Saururaceae
Parts Used: whole plant
Season: summer, fall
Region: Southwest, Mountain West,
　　Pacific Northwest

To gather their medicinals and remedies, most native herbalists must go on several collecting journeys, at different times of the year. Many return to the same locations for specific herbs or groups of plants. Some locations are even acknowledged by other herbalists as "belonging" to a specific healer, family, or clan, and will not be approached. However, there are some plants that are so special and widely referred to that herbalists will maintain a

Since it has so many medicinal uses, many indigenous herbalists cultivate yerba mansa, here in pristine bloom, to ensure a constant supply.

stand right out their back door, in a nearby community garden, or—in the case of yerba mansa—around a local spring. I knew one such herbalist from my tribe. She collected water for drinking and cooking from a little flowing spring located in the back end of a shaded hillside *rincon* ("corner"). The bubbling spring created a small pool. In the wet ground around the spring she cultivated several useful herbs, including yerba mansa. Native herbalists around the western states are very familiar with this plant and its high medicinal efficacy. They keep it handy.

USES

I keep a tincture of yerba mansa root around to use for sore throats and colds. Many native peoples make an infusion of the roots and leaves as a cold remedy, to relieve chest congestion and coughs. Often, an infusion of only the roots is used for coughs and sore throats. The roots can also be chewed, and the resulting juice swallowed to treat coughs and sore throats. The Shoshone take a decoction of the roots for stomach problems and as an antiseptic. An infusion of the entire plant is used as an anticonvulsive. The Pima take a decoction of the crushed root as an emetic and as a laxative. A powder of the leaves and root can be applied to wounds and open sores. The powdered root can also be made into a strong tea and used as a mouthwash to help heal gum infections. The Yaqui use the plant for stomach ulcers. On the West Coast, some native peoples use an infusion of the roots to help with menstrual cramps and as a general pain remedy. Some people, such as at Isleta Pueblo, collect the seeds of yerba mansa, grind them into a flour, and add it to cornmeal or mesquite flour for making breads.

IDENTIFICATION AND HARVEST

Yerba mansa is an herbaceous perennial with a thick woody rhizome. The simple, alternate leaves of the plant release a spicy aromatic smell when crushed. Each plant produces what appear to be white flowers, occasionally with a little reddish tinge. Actually the inch-long flower "petals" are bracts that surround a single inch-long conical inflorescence, which houses about 100 tiny flowers. The flowers bloom all spring and summer. By late fall, only the rust-colored flower stalks and green leaves remain. Yerba mansa inhabits several plant communities, such as valley grasslands, saltgrass flats, and desert fan palm oases. Look for it in saline, damp and wet places, ponds, and bogs from northern Mexico and into the Southwest, along the front range of New Mexico and Colorado, west to the Mojave Desert and coastal ranges, and north to Oregon; occasionally, plants are found in Utah, western Texas, Kansas, and Oklahoma. Yerba mansa does not do well in elevations above 6,000 feet. Be sure to collect only from healthy, vigorous plants and to leave some specimens in the area for future growth. If you would like to start your own stand, close to home, you should carefully collect the seeds in early fall, before the flower spike dries and drops from the plant.

HEALTH BENEFITS

Yerba mansa contains sesamin and asarinin, furofuran lignans that are effective against nontuberculous mycobacteria. The roots contain thymol, methyleugenol, and piperitone. Thymol is an antibacterial and antifungal. Methyleugenol is a carminative. Piperitone and another essential oil demonstrate potential anticancer activity.

YUCCA
YUCCA SPP.

Family: Agavaceae
Parts Used: whole plant
Season: year-round
Region: North America

Sometime around the age of 13, a traditionally raised Apache girl takes part in a four-day-long ceremony that marks the end of childhood and her entry into womanhood. Over the course of the four days, she steps across several physical, emotional, and spiritual thresholds. As part of the ritual, her hair is ceremonially cleansed with yucca root. I attended a portion of one of these ceremonies at the White Mountain Apache reservation in eastern Arizona; the scene was otherworldly. Through this ritual, the young girls become Changing Woman, a figure from Apache lore who was partly responsible for bringing the Apache into existence. For those four days, each initiate assumes the healing and other powers of Changing Woman. She

Yucca baccata features an impressive flower stem, up to 40 inches tall.

becomes the finest example of Apache womanhood. And she can only be cleansed with the root of a plant that many people rarely give a second thought to.

USES

Most of the approximately 27 *Yucca* species native to North America occur on the Great Plains and parts west. For this entry I will focus on three yuccas of the greater Southwest, banana yucca (*Y. baccata*), Navajo yucca (*Y. baileyi* var. *navajoa*), and narrowleaf yucca (*Y. angustissima*). Yucca is a utilitarian plant. It is used in indigenous ceremony and boasts several medicinal qualities; parts of the plant are edible, and other parts can be made into tools and even clothing.

Apache, Navajo, and Pueblo peoples use the root for ceremonial hair washing and in naming ceremonies and other rituals where purification is needed. The root is harvested and, while still fresh, is cleaned and pounded into a pulp-like consistency. The pounded root is infused in warm water and, when the occasion demands it, is vigorously rubbed, until a soap-like froth is attained. This yucca root infusion is then immediately applied to the hair and scalp for washing, which has two effects: the first sensation is an intense tingling in the scalp; the second result is actual squeaky-clean hair. Juice from the thicker leaves of banana yucca provides Navajo midwives with a lubricant they frequently employ to ease childbirth. The roots and leaves of banana yucca, Navajo yucca, and narrowleaf yucca are pulverized and made into a poultice to treat sprains and arthritis in the joints. The young white-green flowers can be eaten raw, steamed, or boiled. The fruit pods that eventually ripen from the flowers can be roasted and eaten.

Later still, the seeds from the ripened and dried fruits are roasted and pounded into edible cakes. The long fresh leaves of yucca are easily bent and folded into temporary footwear. The same leaves can be carefully pounded until the pulp is separated from the long fibers. The

Yucca pods emerge once the plant has finished flowering.

fibers themselves can then be twisted into string, and the string later twisted and braided into rope material. Some yuccas (e.g., banana yucca and Navajo yucca) often sprout a single flowering stalk that can reach lengths of 6 to 7 feet. I speak from experience: the stalk makes a durable, lightweight walking stick as well as a pole for an emergency lean-to shelter during a thunderstorm.

The seeds that are housed inside yucca pods can be roasted and pounded into a flour that's made into edible cakes.

IDENTIFICATION AND HARVEST

Yuccas are easy to identify by their basal rosettes of narrow, closely spaced leaves. The edges of the thin leaves are very sharp and often have short, white, curling hairs. Across the Great Plains and the West, yuccas are found in mixed environments, including deserts, grasslands, mountains, and even coastal scrub in southern and central California. In dry conditions, the plants tend to remain small and grow as individuals; however, in wetter environments, all species of yucca can grow tall and branch. The flowers are white, sometimes with a greenish tint; they are bell-shaped and are usually borne in massed clusters on a central stalk. The stalks and flowers may not develop in drier conditions but will emerge at least once a year if weather conditions are favorable. Yucca can be harvested throughout the year. Collecting the leaves must be done with care, as they are sharp. Use a long machete-like tool and leather gloves to avoid being jabbed and cut. If you intend to collect the root, be prepared to use a good pickaxe and shovel. The roots often grow deep into hard-packed rocky soil.

HEALTH BENEFITS

Yucca root contains silica and steroidal saponins, the latter of which likely accounts for the anti-inflammatory and soap-creating effects of the root. The silica provides the stinging sensation that occurs during the washing. Yucca root and bark also contain yuccaols, yuccaone, and larixinol, phenolic constituents with potential antioxidant effects.

Glossary

ABORTIFACIENT	induces or brings on menstruation or a miscarriage
ALTERATIVE	restores the proper function and balance of the body
ANALGESIC	decreases the body's response to pain
ANDROGENIC	reduces the absorption of testosterone and dihydrotestosterone
ANTHELMINTIC	rids the body of parasites
ANTIANGIOGENIC	prevents tumors from growing their own blood cells
ANTIARTHRITIC	prevents and relieves symptoms of arthritis
ANTIBACTERIAL	seeks out and kills (or inhibits the growth of) bacteria
ANTIBIOTIC	inhibits the growth of microorganisms
ANTICANCER	demonstrates activity against malignant diseases

ANTICARCINOGENIC	counteracts carcinogenic effects and inhibits cancerous growths
ANTICOAGULANT	retards or inhibits the coagulation of blood
ANTICONVULSANT	inhibits convulsive seizures
ANTIDIABETIC	helps to regulate diabetes by lowering blood glucose levels
ANTIFUNGAL	prevents or inhibits the growth of fungal organisms
ANTIHYPERTENSIVE	reduces high blood pressure, preventing strokes and heart attacks
ANTIHYPERURICEMIC	prevents the overproduction of uric acid
ANTI-INFLAMMATORY	minimizes the body's response to injury by reducing pain and the release of neuro-chemicals to the injury site
ANTILITHIATIC	prevents the formation of kidney stones
ANTIMICROBIAL	kills or stops the growth of microorganisms
ANTIMUTAGENIC	reduces the rate of cell mutation
ANTIOXIDANT	protects cells against damage from harmful free radicals
ANTIPLEURITIC	decreases inflammation in the tissue surrounding the lungs
ANTIPROLIFERATIVE	restricts cancerous cell growth in tumors
ANTIPYRETIC	prevents and reduces fever
ANTIRHEUMATIC	slows the progression of rheumatoid arthritis
ANTISEPTIC	retards the growth of microorganisms
ANTISPASMODIC	reduces muscle spasms
ANTITUSSIVE	suppresses coughing

ANTIULCER	prevents the progression of ulcers
ANTIVIRAL	prevents and inhibits viruses
ANXIOLYTIC	relieves anxiety
ASTRINGENT	encourages the contraction of skin cells and body tissues
CARMINATIVE	prevents or relieves flatulence
CATHARTIC	a purgative herbal therapy
COMPOUND	the metabolic chemicals of plants (e.g., phytosterols, lipids, essential oils)
CONSTITUENT	the medicinal phytochemicals of plants
CYTOTOXIC	toxic to living cells
DEMULCENT	helps to relieve inflammation
DERMATOLOGICAL AID	helps skin heal from wounds, cuts, rashes, and dermatitis
DIAPHORETIC	helps to increase perspiration
DIURETIC	promotes the production of urine
EMMENAGOGUE	stimulates and increases menstrual flow
EXPECTORANT	promotes the secretion of mucus and saliva in the air passages
FEBRIFUGE	reduces fever
GASTROPROTECTIVE	protects the stomach lining from irritation
HEMOSTATIC	helps to prevent or stop bleeding
HEPATOPROTECTIVE	helps to prevent liver damage
HEPATOTOXIC	associated with liver damage
IMMUNOMODULATORY ACTIVITY	stimulates antibodies to strengthen the immune system

LARVICIDAL	targets the larval stage of an insect
LAXATIVE	helps to loosen stools and promote bowel movements
PHYTOCHEMICAL	biologically active plant chemicals
PHARMACOLOGICAL	effects and actions of plant chemicals on and in the body
PHARMACOLOGICALLY ACTIVE	plant constituents that are metabolized and demonstrate effects on and in the body
PHARMACOTHERAPEUTIC	beneficial effects of plant constituents
STOMACHIC	promotes appetite and helps with digestion
STYPTIC	stops bleeding
SUBTONIC	retards growth
URICOSURIC	increases the secretion of uric acid in the urine
VASODEPRESSOR	promotes the loss of consciousness
VASODILATOR	helps to widen blood vessels

Further Reading

Curtin, L. S. M. 1997. *Healing Herbs of the Upper Rio Grande*. Santa Fe: Western Edge Press.

Dunmire, William W., and Gail D. Tierney. 1995. *Wild Plants of the Pueblo Province*. Santa Fe: Museum of New Mexico Press.

———. 1997. *Wild Plants and Native Peoples of the Four Corners*. Santa Fe: Museum of New Mexico Press.

Goodrich, J., C. Lawson, and V. P. Lawson. 1996. *Kashaya Pomo Plants*. Berkeley, California: Heyday Books.

King, Thomas. 2003. *The Truth About Stories*. Toronto: House of Anansi Press.

Moerman, Daniel E. 1998. *Native American Ethnobotany*. Portland, Oregon: Timber Press.

———. 2009. *Native American Medicinal Plants*. Portland, Oregon: Timber Press.

Timbrook, Jan. 2007. *Chumash Ethnobotany*. Santa Barbara Museum of Natural History Monographs. Berkeley, California: Heyday Books.

Turner, Nancy J. 2014. *Ancient Pathways, Ancestral Knowledge*. Montreal: McGill-Queen's University Press.

Willard, Terry. 1992. *Edible and Medicinal Plants of the Rocky Mountains and Neighbouring Territories*. Calgary: Wild Rose College of Natural Healing.

Acknowledgments

Very few events in our universe occur in isolation. An action will lead to a reaction and then to interactions, which lead to complex stochastic systems and random probabilities. This book has evolved in much the same manner. The action of my writing would never have occurred if not for an email from Timber Press editor Stacee Lawrence. Stacee contacted me wondering if I might be interested in writing a book about American Indian ethnobotany. Her confidence in my abilities and steady and unobtrusive support have been important to the completion of this work. After excitedly agreeing to write this book, I began to make contact with my network of ethnobotanical friends and colleagues, among them Nancy Turner, Phyllis Hogan, Peter Forbes, and Julie Pyatt, who made important and welcomed suggestions regarding which plants to include in this compilation. My interactions with these wonderful people led to further sources of plant-related ideas and exchanges with additional experts, including Judy Dow and Roy Upton. Thanks to Bob and Audrey Erb for the photos of their birch bark canoe. Finally, I am grateful to my wife, Lisa, who afforded unending support for this project and worked tirelessly to stymie the barking and interruptions of our three dogs while I tried to write.

Photography Credits

Michael A. Dirr, page 26.

National Museum of the American Indian, Smithsonian Institution, pages 27 (17/6240); 109 (7/2269); 173 (right) (8/9757); 176 (bottom) (25/9282); 178 (top right) (2/2225). Photos by NMAI Photo Services.

Biodiversity Heritage Library, pages 28, The North American sylva; or, A description of the forest trees of the United States, Canada, and Nova Scotia . . . vol. 3, plate 118 / Michaux, François André, and Smith, J. Jay; 36 (right), Köhler's Medizinal-Pflanzen . . . vol. 3, plate 69 / Brandt, Wilhelm, and Gürke, M., and Köhler, F. E. (Franz Eugen), and Pabst, G. (Gustav), and Schellenberg, G. (Gustav), and Vogtherr, Max; 41 (right), Medicinal plants: Being descriptions . . . vol. 3, page 196 / Bentley, Robert, and Trimen, Henry; 55 (top), Österreichs allgemeine baumzucht . . . vol. 3, plate 161 / Franz Schmidt; 100 (right), Curtis's botanical magazine, vol. 29-30, no. 1148-1236, plate 1198.

Getty Images, page 29 (right), DEA / S. AMANTINI / Contributor.

Adobe Stock, pages 31, 32, USantos; 35, 126 (left), clubhousearts; 38, 141, 196, Andrei; 39, maxandrew; 41 (left), IKvyatkovskaya; 44, bjphotographs; 45, nickkurzenko; 48, Andris Tkachenko; 53, photodigitaal.nl; 61, spetenfia; 70, Tina_Jeans; 72, John; 73 (top), Serghei Velusceac; 73 (bottom), George Sheldon; 75, Mieke Vleeracker; 78, 133, The Nature Guy; 80 (top), karlo54; 82, Viktor; 83, ajlatan; 89, Irene Teesalu; 94, Reimar; 97, alenazamotaeva; 106, airdenet; 112, Mary Brenner; 113, Stephen Bonk; 114 (right), Ron Rowan; 115 (left), Hank Erdmann; 119, Rob Mutch Photo; 120 (top), Joy Fera; 121, Nikki; 126 (right), Le Do; 130, Keith; 131, Volodymyr; 132, akhug; 134, helga_sm; 135 (left), JOW; 146, iredding01; 148, fotocinema; 151 (right), monamakela.com; 153, Haramis Kalfar; 156, PhotoSpirit; 157, Teressa L. Jackson; 158 (top), aifeati; 163 (left), Sue Smith; 170, qingwa; 172, 173 (left), Dennis; 174, JLV Image Works; 175, leemarusa; 176 (top), Jackie DeBusk; 177 (left), nicemyphoto; 178 (left), hakat; 179, tarttong; 184, Didier SCUVIE; 187, argenlant; 188, Virtexie; 189, Grigory Bruev;

190, Designer; 191, Vitaly Ilyasov; 194 (top left), Starover Sibiriak; 194 (bottom left), Henri Koskinen; 195, Krzysztof; 198, simona; 199 (left), Mannaggia; 203 (left), Mark Herreid; 205 (right), photopic; 207, imfotograf; 208 (right), agneskantaruk; 210 (left), oxxyzay; 211, Maksym Dragunov; 214 (left), Veronika; 215, Shakzu; 217, Laurens.

Audrey Erb, pages 34, 208 (left).

National Museum of Natural History, Department of Anthropology, Smithsonian Institution, pages 36 (left) (E260522-0); 90 (E200721-0).

Wikimedia Commons, pages 37, Famartin; 42, Daniel Case; 66, Walter Siegmund; 108, SB Johnny; 124, Samartur; 150, JerryFriedman; 165 (left and right), Kenraiz.

University of Northern British Columbia, page 43.

Eugene Sturla, courtesy of Southwest Desert Flora, pages 47, 139.

©Nancy Bundt / Photographer, page 49.

Alamy, pages 50, 104, Gerry Bishop; 51, Dorling Kindersley ltd; 55 (bottom), blickwinkel; 79, Kevin Knight; 80 (bottom), 160, imageBROKER; 84, Ken Barber; 93, Loop Images Ltd; 99, Steffen Hauser / botanikfoto; 105 (top), Mindy Fawver; 117, Bob Gibbons; 136, 168, Robert Shantz; 140, Robert Mutch; 147, Avalon / Photoshot License; 178 (bottom right), Album; 193 (bottom), Dawna Moore; 201 (left), Andrei Stanescu; 203 (right), Tim Gainey; 213, Hotaik Sung; 214 (right), Judy Freilicher.

Rachel Mackow, page 52.

Margo Bors, page 56 (left).

Shades of Rez Studio, "Mariposa Musings" flute designed and created by Tim Blueflint Ramel, page 57.

Flickr, pages 60, ClatieK; 63, Tim Sheerman-Chase; 102, Katja Schulz; 110, Wings in the Park; 125 (bottom), veggiefrog; 152, JOE BLOWE; 166, Matt Lavin; 167 (right), InAweofGod'sCreation; 205 (left), Eric Toensmeier; 210 (right), Keith A. Bradley.

Shutterstock, pages 62, tamu1500; 65, Peter Turner Photography; 67, Gatis Grinbergs; 100 (left), Danny Hummel; 107, vaivirga; 125 (top), Serghei Starus; 127, EasyArts; 129, You Touch Pix of EuToch; 161 (top), javarman; 167 (left), ttoleg; 183, Playa del Carmen; 185, E.J.Johnson Photography; 194 (right), bjphotographs; 218, rSnapshotPhotos.

Library of Congress, Prints & Photographs Division, pages 64, Edward S. Curtis Collection (LC-USZ62-108464); 77, Miscellaneous Items in High Demand Collection (LC-USZ62-133885); 95 (bottom), Edward S. Curtis Collection (LC-USZ62-118586); 145, Edward S. Curtis Collection (LC-USZ62-103072); 154, National Photo Company Collection (LC-DIG-ds-12610); 158 (bottom), FSA/OWI Collection (LC-USF34-035945-D); 201 (right), Edward S. Curtis Collection (LC-USZ62-98673).

Carving by Benjamin Kabinto, courtesy of Kachina House, page 74.

Lady Bird Johnson Wildflower Center, pages 86, Alan Cressler; 87, Will Stuart; 103, Julie Makin.

Maryland Biodiversity Project, pages 88 (left), Ashley M. Bradford; 88 (right), Kirsten Johnson; 101, Wayne Longbottom; 177 (right), Jim Brighton.

Index

About the Author

Lisa Gaspari-Salmón

Enrique Salmón is a Rarámuri (Tarahu-mara). He is head of the American Indian Studies Program at Cal State University–East Bay, in Hayward, California. He holds a BS from Western New Mexico University, an MAT in Southwest studies from Colorado College, and a PhD in anthropology from Ari-zona State University. He has been a scholar in residence at the Heard Museum and has served as a board member for the Society of Ethnobiology. He has published many arti-cles on indigenous ethnobotany, agriculture, nutrition, and traditional ecological knowl-edge. He has also spoken at numerous confer-ences and symposia on the topics of cultivating resilience, indigenous solutions to climate change, the ethnobotany of Native North America, the ethnobotany of the Greater Southwest, poisonous plants that heal, bioculturally diverse regions as refuges of hope and resilience, and the language and library of indigenous cul-tural knowledge.